THE NATURE OF SCIENCE

An Epistemological Analysis

V. Ilyvin

A. Kalinkin

First LG Edition 2018

ISBN 978–93–83723–42–3

Published by
LG PUBLISHERS DISTRIBUTORS
49, Street No. 14, Pratap Nagar,
Mayur Vihar Phase I, Delhi 110091
lgpdist@gmail.com

Printed at
Sapra Brothers, Noida

CONTENTS

PREFACE

What is the nature of science as a categorial-logical relation to the world as distinct from the other products of spiritual-theoretical assimilation of reality? What is science conditioned by? Where lies the boundary between science and the sphere of everyday experience, artistic thinking, morality, religion, and other areas of the social superstructure?

Throughout history, epistemologists have attempted to answer these questions—to create an acceptable model of the scientificity of knowledge. These attempts, however, have, we believe, failed due to the inadequacy of the philosophical-methodological foundations of the concepts of science developed in pre- and non-Marxian epistemology, which raised to an absolute either the empirical and inductive, or the rational and deductive, or else the subconscious and intuitive aspects and components of knowledge. These models of science were therefore narrow and non-historical; they ignored the social status of cognitive activity and negated practice as the final instrument of ascertaining the scientificity of knowledge. These difficulties were only overcome in dialectical materialism, which provided a truly scientific foundation of epistemology, as well as a worldview basis adequate to the complexity of the problem.

Being a manifold phenomenon, science is studied in the framework of approaches based on logic and methodology, social psychology, economic statistics, information theory, cybernetics, etc. The present work, taking the epistemological approach, is focused on the gnoseological problems of science as knowledge. Accordingly, other questions which, though associated with this problem range conceptually, thematically, etc., merit special consideration, are excluded from the universe of discourse.

There are various kinds of epistemological units or taxons: theory, family of theories, paradigm, research programme, intellectual tradition, types of knowledge, finally, knowledge as a whole. The principal unit of analysis accepted here is scientific knowledge and its forms, kinds and types.

The view of science as knowledge is important not only in that it expands the boundaries of the traditional logical-methodological analysis to encompass, along with the fairly well-studied "theory", the relatively little-studied kinds of knowledge and knowledge as a whole, but also in that it opens up fresh perspectives for constructing a general epistemological theory of science as a specific type of rationality, a special kind of intellectual production.

The structure of the work is determined by the mode of research adopted in it. First, general problems, such as the criteria of scientificity and science as knowledge, are studied. This is followed by a discussion of less general but epistemologically highly important problems of science: the genesis of science, the definiteness of the forms of scientific knowledge, and the dialectics of the development of science. The choice of these problems for analysis was determined by the fact that not all of the aspects of this problematics have, in our view, been sufficiently studied in scientific literature.

The book consists of four chapters. Chapters 1, 2 and 4, as well as the Preface and Afterword, were written jointly by the two authors; Sections 1 and 2 of Chapter 3 were written by Viktor Ilyin, Sections 3 and 4 of that chapter, by Anatoli Kalinkin.

Chapter I.

THE CONCEPT OF SCIENCE

Philosophers have always faced the task of studying the dialectics of scientific knowledge and its epistemological foundations. The focus here is largely on bringing out the epistemological nature of scientific knowledge as opposed to non-scientific forms of knowledge. We shall begin our discussion of this topic with the problem of criteria for determining that nature.

1. 1. CRITERIA OF SCIENTIFICITY

The actual multidimensionality and plurality of the strata of intellectual production give rise to the problem of specification of its constituents. Accordingly, theories of mythological, artistic, religious, and practical everyday experience come into being. They are concerned with defining the specific features of both the human activity and its products in each concrete case. The theory of science figures prominently in this series of theories. Its task is reflexion on the basis for drawing a demarcation line between science and non-science. (Generally speaking, the sphere of non-science is wide and heterogeneous, including as it does non-scientific forms of cognitive activity within the framework of practical—everyday, artistic, etc.—experiences; pre-science, or proto-knowledge—the basis of future science; pseudoscience—the fantasies and prejudices masquerading as science (e.g., phrenology); para-science, or knowledge whose epistemological status does not satisfy the conditions of science, such as parapsychology; and anti-science—deliberate distortions of the scientific view of the world, as found, e.g., in the bourgeois social utopias in sociology.)

What are the epistemological statutes of science? What distinguishes the scientific approach from other types of attitude to the world? These and similar questions can only be solved through identifying certain objective indicators or criteria of scientificity. The problem of establishing such indicators is, in fact, the problem of the criteria of scientificity.

If we accept the traditional interpretation of criterion as a rule for making choices and realising preferences the following definition of criteria of scientificity can be suggested. Criteria of scientificity are rules for evaluating the products of cognition as according, or not according, with the standards of science; they permit an ordering of the products of cognition in terms of their closeness to (or remoteness from) science; they are fundamental theoretical-methodological principles, norms, values, ideals, or standards conditioning the definiteness of the foundations in accordance with which tentative knowledge (a corpus of ideas—hypotheses, concepts, theories, assumptions, or facts) and tentative activity (a corpus of acts—thinking, theorising, conceptualisation, or experimenting) are assessed as knowledge and included in the category of science.

Criteria of scientificity thus provide a normative and axiological orientation for research, direct research activity, reject non-productive orientations, establish model methods for the generation of knowledge, and point the directions of desirable evolution of various branches and disciplines, selecting the units of knowledge on the basis of preference filters available in cognition.

Criteria of scientificity are norms; their definiteness, just as the definiteness of any other norms, is specified by the dispositions, sanctions and conditions, of action.

Dispositions are sets of directions, instructions, recommendations, imperatives, interdictions, etc., which characterise a mode of cognition. As concentrations of demands imposed on knowledge and activity, dispositions fall into two mutually complementary segments. One of these includes an ensemble of rules of the "what to do" type: "seek for sufficient reasons", "minimise argument", "exclude formal contradictions", etc. These may be referred to as positive heuristics. It increases the chances of obtaining epistemologically desirable products of research. The other set comprises rules of the "what not to do" type: "do not introduce *ad hoc* hypotheses", "do not follow authorities blindly", etc. That is negative heuristics. It employs a system of interdictions in order to bar from science ideas and mental moves known to lead into blind alleys.

Sanctions assure the effectiveness of dispositions. The point is that criteria of scientificity embody the most promising and fruitful principles and modes of action in science, expressing the objective laws and the logic of its historical change. By specifying the concepts of acceptability, desirability, and preferability of some kinds of cognitive assimilation of reality, and those of untenability, unacceptability, and defectiveness of others, criteria of scientificity mould the axiological and normative self-consciousness of science. Criteria of scientificity thus perform protective functions, guarding science against exotic (in a negative sense) ideas and creative initiatives. In science, for instance, it is impermissible to violate the laws of conservation. Although that does not at all mean that these laws cannot be modified or revised, any research laying claims to a scientific status must, for the time being at least, satisfy the demands of these laws.

As an integral object rather than an agglomeration of qualitatively different phenomena, science is supported by an ensemble of invariant principles. The content and meaning of these principles are naturally relative, but the requirement itself of the existence of such principles is absolute. The rules or instructions of the dispositions rank precisely as such principles.

What happens if dispositions are ignored or deformed? In this case, sanctions come into effect.

The question of ignoring dispositions appears to us to be trivial. That which contravenes the demands of science and is posited out of ignorance (things like fabrications and prejudices) is unscientific in its very essence. The question of deformation of dispositions appears to us to be nontrivial. At the beginning, that which contravenes existing science is also declared to be unscientific; yet it may be essentially scientific (cf. scientific revolutions restructuring the systems of available criteria of scientificity). Putting off a discussion of the dynamics of criteria of scientificity to a later stage, let us here stress the role of sanctions. Playing a somewhat conservative function, sanctions protect science against trivially extravagant ideas, they set an optimal operational mode of research, guarantee a minimum of risk in the search for the best possible results, and ensure the continuity and balance of scientific experience.

The conditions of the functioning of norms set down the features of situations that are possible in science, they specify the requirements of the criteria of scientificity in relation to the various kinds of knowledge and activity, and lend meaning to the testing of the correspondence between the multiform complexes and components of science and an extensive network of special and general criteria of scientificity.

Criteria of scientificity are varied and multilayered. They are subdivided into three sets. The first of these (which we shall designate "set A") comprises universal criteria of scientificity which draw the demarcation line between science and non-science. They set down the limits, as it were, of the basis on which the concept of unitary science is constituted regardless of its differentiation in terms of subject matter, method, and professional forms. Such norms figure here as formal consistency, cause-and-effect cohesion, experimental verifiability, rationality, reproducibility, intersubjectivity, etc. The requirements comprised in set A are necessary ones. For instance, rationally (or logically, practically) unsubstantiated elements have no place in science—otherwise we are faced with a move beyond its boundaries. The criteria of scientificity of set A comprise those inviolable schemata, archetypes, and principles of the intellect which, in their positiveness, condition the identity and integrality of science as a stable internally organised phenomenon. They make science a unified and synchronically active system embracing autonomous yet structurally similar, epistemologically isomorphic kinds of knowledge.

The second set, which we shall designate "set B", is a group of historically transient normatives which determine the process of modelling, simulation, explication, interpretation, and meaning formation; they make the course of events and the order in nature intelligible and specify rational samples of connections in terms of which the course of events can be meaningfully discussed. Here belong such normatives as the requirements imposed on ontological schemata, existence hypotheses, epistemological assumptions, pictures of the world, and so on, which are oriented toward research programmes, ideals of knowledge, etc., accepted in various cultures. Unlike those of set A, the criteria of scientificity of set B merely specify the cultural-stylistic dimensions of the thinking of scientists, being of fundamental significance for characterising knowledge in its concrete historical projection.

The third set, to be designated "set C", is a group of subject-matter criteria of scientificity imposed on professionally differentiated (in terms of subject matter and method) branches—the various systems of knowledge and activity. Requirements specific for logico-mathematical, natural, and technical sciences, as well as for the separate genera and modifications of knowledge, concrete theories, hypotheses, cognitive acts and their complexes, are identified here. As opposed to the first two sets, set B criteria are fairly narrow, being instruments of characterising the concrete kinds of knowledge and activity and reflecting particular rather than typical parameters of science.

The genera of criteria of scientificity specified here permit the modelling of scientificity as a multi-layered entity having a certain nucleus (criteria of scientificity of set A) as well as a historical (set B) and subject-matter, thematic (set C) dimensions. However mobile and polymorphous, science thus proves to be exhausted, in a sense; there is nothing in the world of science that would not be covered by the system of criteria of scientificity.

Is scientificity expressed by all the criteria taken together? In a certain respect, yes; but that respect is extremely abstract and highly idealised. Science appears here as too perfect a knowledge—fully adequate, and devoid of any defects, gaps, inconsistencies, etc. The actual situation is quite different. Since science includes, along with perfect elements, elements of the imperfect, without which science is simply nonexistent, the idealising interpretation of scientificity, though quite useful in some respects (introducing as it does the world of what must be, a world that has a regulative and goal-setting meaning for the scientist), must be revised in the sense of bringing it closer to and in agreement with the world of what is. The revision comes from the realisation of the differentiation within the body or the ontology of science. The latter is made up of frontline science, the hard core of science, and the history of science.

A minimal condition of incorporating some result in frontline science is the obtaining of it by scientific means. The result may be true or false, sufficiently or insufficiently substantiated (this will become clear *post festum*), but it must fall within the framework established by the rules and standards of obtaining results in science as a sphere of the social superstructure. The hard core of science is the segment formed entirely by true elements absorbed from the total body of science. This is, as it were, the sedimental base, the evidential basis of science—a reliable layer of knowledge crystallised in the course of its progress. The history of science is a fragment formed by the body of knowledge thrust beyond actual science and morally outdated. On the methodological approach, however, it would be meaningless to oppose this fragment to science as such. Science is dynamic, processual, and historical. What scientists do now will in future become child's play; the character, content and meaning of science change, but results obtained at one time according to the rules of science do not lose the feature of scientificity. Besides, it is important to see the intransient value of this fund. It is, in fact, an eternal preserve of ideas: something realised in the past can be reanimated it the future. Of this nature is, say, the idea of atomism, which has invariably provided sustenance for knowledge.

The distinction between these segments of science is fruitful in two respects. Firstly, it stresses once again the limited meaning of the idealising interpretation of scientificity. Not all knowledge that forms science is perfect—if we have in mind the realisability of such strong criteria as truth, sufficient substantiation, etc. Frontline science contains unidentified unsubstantiated and false elements; apart from that which will become part of future science, it comprises that which will remain outside science. The history of science contains identified unsubstantiated and false elements—the ballast that failed to make the transition from frontline science to the hard core. Secondly, it lends a graphic quality to the assertion of the ontologically multi-layered structure of science, which, together with the realisation of the functions or purpose of each part integrated in science as a whole makes it necessary to describe or represent them (and through them, the manifold reality of science) in sovereign constructs—in specific values and normatives. The latter necessitates a differentiated picture of scientificity.

In frontline science, such regulative figures as nontriviality, informativeness, heuristic quality, etc., are focal. At the same time the requirements of exactness, rigorousness, substantiatedness, etc. are weakened and made less radical. The reason is that the purpose of frontline science is to vary the alternatives, to go through all the possibilities, to extend semantic horizons, and to produce the new. If the requirements of exactness, rigorousness, substantiation, etc., were to be imposed on all the components of science from the outset, science would be a collection of trivia. To some extent, informativeness and rigorous substantiation rule each other out: in frontline science, the falsity of an informative contradiction is preferable to the truth of a truism. Science must contain rigorously substantiated elements, but not only such elements; otherwise science will lose its heuristic quality. For this reason, badly substantiated elements (insufficiently confirmed, thoroughly refuted, "crazy" ideas, etc.) must be admitted to science, but these must not exhaust science.

Frontline science is the most hypothetical, problematic and unreliable segment of science; it includes a corpus of probable and not very probable knowledge which is not, however, rejected. It is not rejected for many reasons, such as (1) its inconsistency is not proved; the absence of such proofs is in itself an argument in favour of accepting this hypothetical knowledge; (2) there are hopes for its future substantiation; (3) critical verification of hardly probable (hypothetical) knowledge catalyses the production of new knowledge; elimination of errors and deviations from the body of science would deprive it of ability for progress; (4) proliferation of hypothetical knowledge reduces the probability of leaving essential possibilities unutilised; the need for selecting theoretical alternatives increases the flexibility, dynamism, critical spirit, and conclusiveness of science.

Thus the importance of this segment of science is not connected with the plurality of truths, as Feyerabend believes, but with the fact that the path which future approximation of the truth will take is unknown. Many elements of frontline science will be rejected, but their incorporation in the totality of science is justified because, first, they are obtained by scientific means, and second, the mode of establishing their falsity helps to find the direction of new inquiries and to build new theories.

The hard core of science is regulated by such values and norms as clarity, rigorousness, reliability, substantiatedness, conclusiveness, etc. The task of frontline science is to generate new knowledge. It is therefore woven of the vicissitudes of the search for the truth—of presentiments, wanderings, separate breakthroughs towards clarity, etc. Its knowledge is therefore minimally reliable. The task of the hard core of science is to be a definiteness factor, a source of premises and of basic knowledge which orients and corrects cognitive acts.[1] It therefore consists of proofs and substantiations, embodying the most stable and objective part of science. Extraordinary grounds are needed for its modification or critique.

This interpretation is in fine agreement with Klein's idea of the existence of two periods in the development of knowledge. One period, that of irrepressible growth of creative productivity, is identifiable with frontline science. The other period, that of sifting and purification of the achievements, can be identified with the hard core of science.

Of course, our interpretation of the hard core, just as, incidentally, the interpretation of frontline science, should not be taken literally. Its truth is evinced in the tendency. In actual fact the body of the ideally demonstrable is not very great in the hard core—it only includes a few propositions of logic and finite arithmetic, which are also problematised [consider, e.g., the law of excluded middle (*tertium non datur*) and others].

The history of science is a fragment of science first and history second. It would be erroneous to view historico-scientific activity as archival or archaeological activity only, limited to the search for and processing, systematisation, etc., of facts relating to the past of science. Historico-scientific activity is in the first place a scholarly activity; it forms part of research. The history of science stimulates research (offering such model programmes of research as elementarism, evolutionism, etc.); it contains a comprehensive panorama of the dynamics of knowledge, facilitating the perception of intrascientific perspectives and possibilities (whence, how, where to, why, and what for); accumulating information about the ways of achieving knowledge, about the forms and modes of analysing objects, it performs educational functions.

All these aspects of the history of science promote an adequate cognitive reconstruction of the object under study, which makes it an integral part of science.

Different parts of science are apparently dominated by different values and norms.

The hard core of science, along with the history of science, works as an instrument of eliminating all kinds of extravagancies. Acts of selection are based on the criterion of agreement, which reads: that unit of knowledge (a hypothesis, a theory) is better which is in better agreement with the criteria of sets B and C—the basic knowledge of the hard core and of the history of science. But that is by far not all. The proposition concerning the relative immutability of set B criteria and the need to bring fresh knowledge introduced into science in agreement with these criteria has a great heuristic potential. In some cases, the testing of candidates for a place in science for agreement with set B criteria, composed of fundamental laws, leads to discoveries.

Thus when energy leakage in β-decay was discovered, Bohr proposed to sacrifice the law of conservation and transformation of energy (in the case of the microworld). This idea failed, however, precisely because of the fundamental nature of this law. Moreover, relying on this law as an element of fundamental knowledge, Pauli was able to overcome this difficulty by suggesting the neutrino hypothesis which, as we know, was later confirmed. This example, however, is an exception rather than the rule.

The strategy of the hard core of science largely restricts the ability of knowledge for progress. Indeed, the better knowledge is correlated with set B criteria, the less informed and heuristic it is; the progress in knowledge is incompatible with its high probability in respect to B. The growth of the content of knowledge, associated with decreased probability of agreement with set B criteria, is fully ensured in frontline science, where other values and norms are adopted.

Frontline science being an instrument of self-expansion and self-development of knowledge, daring, a tendency towards revising well-established truths, and conflicts with set B criteria are inherent in it. A different logic of evaluating the products of cognition therefore obtains here. "Informative force" and "non-triviality of knowledge" rather than the measure of confirmation are taken into account here. Inasmuch as informativeness is in inverse proportion to maximal probability of knowledge in relation to set B criteria, its internal merits are evaluated in terms of daring and degree of novelty.

Preference is given to that knowledge whose dependence on additional knowledge is greater than that of other fragments of knowledge, and which, after additional knowledge has been proved, is marked by a greater degree of probability.

Thus the problem of scientificity is not solved by an idealising interpretation of knowledge. What is needed is a differentiated approach based on the correlation between scientificity and the specific values dominating the various branches of science. This approach also covers the problem of the dynamics of norms arising from the actual progress of cognition.

Criteria of scientificity are not a priori norms whose highest virtue, in the words of Karl Marx, lies in their suprahistorical character—they are generalisations of cognition and results of its interpretation and perception at definite stages. When the orientation of cognition towards the existing criteria of scientificity ceases to be justified, the need arises for altering them.

The problem of a rational boundary between substantiated and unsubstantiated changes of (or deviations from) existing criteria of scientificity is not simple. It seems to have no universal solution. At the same time it is not the kind of question that can have no solution at all. With a view to working out such a solution, let us consider the following.

In science, rational activity is one that is organised in accordance with the norms of scientific reason. But there is a limit to the rationality of the norms, and of reason constituted by these norms. Inasmuch as this limit is practically revealed by actual cognition itself, the question of the rationality of norms, and of science, is also solved practically; it is practice, the practice of actual cognition, that pronounces the verdict on the rational nature of norms, knowledge and activity in science.

Tactically, norms are a preventive safety belt protecting science against extravagances. They are by no means canonised in this function. The point is that "adopting them as an orientation and a programme of activity always involves an element of methodological risk" (36, 376).* The risk lies in their origins: they are generalisations of past and present, not of future science.

We are thus faced with an interesting situation here. On the one hand, science needs norms (criteria of scientificity), for it does not work by trial and error. On the other hand, it cannot work out from the outset methodological rules that would be adequate to its final results. It has to use rules justified by the past history of science, but their extrapolation to the future is not always justified. Norms programme the activity of the scientist, who is by no means free in his choices.

But the situation is different in frontline science, whose purpose (that of generating the new) determines its ability of functioning as an instrument of transforming norms—nothing new can be produced without deviation from norm.

The ability of frontline science to disrupt the cohesion of the available pool of norms lies in the two-dimensional nature of research activity. We refer to the fact that any intrascientific contribution is inseparable from a methodological one; a metascientific reflexion on the results of innovations necessarily takes place. In the course of that reflexion, the innovations are either incorporated in the pool of available norms or else they are not. In the first case, we deal

* Here and throughout this book the first figure in parentheses indicates the number of a reference at the back of the volume, the second, the page; where a reference is to a multivolume publication, the second figure indicates the volume, the third, the page.

with everyday scientific activity, in the second, with a scientific revolution. The following possibilities come to light here.

(1) There is a tendency to link innovations with traditional criteria of scientificity; cf. energy leakage in ß-decay and the Pauli hypothesis (a positive example); or the adaptation of relativistic ideas to the idea of the ether by Lajos Janossy (a negative example).

(2) There is a tendency to reject innovation-conditioned changes of criteria of scientificity; cf. e.g. the difficulties involved in the evolution of the nonclassical theories in physics (the role of the observer, statistical laws, etc.).

(3) There is a tendency to preserve criteria of scientificity by discrediting innovations (cf. the way genetics was torpedoed by invoking the proposition that "science is an enemy of chance").

(4) Innovations may be accepted and criteria of scientificity transformed; cf. the rejection of the view of dynamic laws as universal when quantum mechanics and genetics were accepted).

When is the breakup of criteria of scientificity completed? When is it no longer expedient to follow the available criteria of scientificity?

Thomas Kuhn actually expressed the view that a clear answer to these questions is impossible. "Though the historian," he wrote, "can always find men .... who were unreasonable to resist for as long as they did, he will not find a point at which resistance becomes illogical or unscientific. At most he may wish to say that the man who continues to resist after his whole profession has been converted has *ipso facto* ceased to be a scientist" (158, 159).

We believe that it is possible to answer these questions. The conflict between frontline science and the hard core with the well-established criteria of scientificity associated with it is over when the innovations as components of frontline science are transformed into hard core, becoming part of the theoretical foundation of science. At this moment, the breakup of norms is completed. A new stable normative domain emerges, and a reinterpretation of the content of science in terms of the new values (or criteria of scientificity) begins.

As of that moment, it is, as a matter of fact, both unreasonable and illogical to oppose the new. He who is incapable of overcoming this barrier places himself, *de facto* and *de jure,* outside science.

The transference of innovations from frontline science

to the hard core, and the modification of criteria of scientificity associated with this, is often linked with the replacement of one generation of scientists by another. It is said that opponents are not converted—they die out. Young people, more receptive to the new, accept changes more readily. We reject this approach in view of its non-epistemological character—although there are some grounds for it. On the epistemological approach, the real mechanism of this transference lies in substantiation and proof. It is true, though, that these are also more readily accepted by carriers of less conservative consciousness, that is, by young people.

* * *

The question of what scientificity is cannot be answered in any unambiguous and definite way. Firstly, scientificity is not defined *ex cathedra* by a few hocus-pocus phrases—its conclusive and reliable definition is practice, an allround and deep-going generalisation of the data of production and cognition. Secondly, it does not have the status of a suprahistorical postulate; it is no permanent label, constant characteristic, or immutable state.

Scientificity is processual and dialectical.

The real dynamism and multidimensionality of scientificity, and the abundance of its essential ramifications make its explication difficult.

Exaggeration of the historicity and mutability of the substantive and normative content of science leads to "catastrophism", to insistence on the incommensurability of the structures and standards of knowledge. This latter feature is characteristic of the descriptivism of the adherents of the historical trend in postpositivism, with its mosaic doctrine of science. Scientificity is here viewed as a factor tied to particular cognitive situations and not amenable to a logical normative description.

An attempt to avoid the shortcomings of descriptivism, implemented as a radical rejection of the "historisation" and "ecologisation" of scientific norms, leads to another extreme—to the apriorism of the "critical rationalists". Scientificity is predestined to the real process of cognition "from above", being closed on itself rather than on knowledge. Critical rationalists drive the spirit of realism from epistemology. We refer in particular to their logico-empirical models of scientificity, which do not, in principle, express the properties of science either at the stage of discovery or at the stage of justification.

Descriptivism vs. apriorism. Can this dilemma be resolved? Some believe it cannot. On these grounds the theory of science, allegedly incapable of providing effective schemata of scientificity, is discredited.

The descriptivism vs. apriorism antinomy can be overcome if the dialectical principles of historicity, concreteness, and differentiated consideration are applied to the analysis of science and the latter is perceived as a highly ramified, rational and complex structure with numerous branches having autonomous complexes of values and normatives. On the basis of the latter, science can be selectively unified without relativising scientific norms or making them unrealistic and suprahistorical.

The unification of science as a whole is the purpose of set A criteria of scientificity. The unification of historical massifs of science is the purpose of set B criteria. And the unification of subject-matter and thematic divisions of science is the purpose of set C criteria. These are horizontal unifications, as it were. In the first case, they consolidate science as opposed to non-science. In the second, they delimitate the phases and stages in the evolution of science; and in the third, they isolate the subject-matter and thematic units of science.

Vertical unifications are based on the fact that science is an apparatus for the generation and standardisation of the truth. The lower threshold of scientificity is in this case specified by the mode of verification (obtaining, formulating, defending) of assertions regulated by the canons of logical, empirical and nonempirical substantiation satisfying the appropriate criteria of scientificity.

At the same time, the entire fullness of the truth is not given at the separate stages of cognition; fictions also form part of science, although that is neither deliberate nor known beforehand. The instrument of identification and elimination of fictions is practice, which establishes the truth of scientifically verifiable results of cognition; practice therefore specifies the upper threshold of scientificity.

Practical verification directs the internal rearrangement of the content of science according to the principle of concentration of the truth in the hard core and thrusting the fictions into the history of science.

The axiological scale—the norms, regulators, and stereotypes of science—take shape on the basis of the hard core, of research methods tested by the course of cognition. Any innovation expanding the available pool of knowledge interacts not only with the latter but also with the adopted scale of values. As a result, shifts occur in the scientist's awareness of his discipline and of its system of values. In this way the actual progress of cognition determines the evolution of the norms of science.

The scale of the values of science is multi-layered, owing to the multi-layered structure of science. Each segment of science is correlated with normative filters and confidence intervals of its own; thus that which works in frontline science does not obtain in well-developed science, etc.

At the same time certain features of constancy and continuity can be found here—something very stable, something that is reproduced even in changes seen as revolutionary ones and unites even externally incommensurable and conflicting elements. These common elements are such fundamental values as "truth", "objective substantiatedness", "rational necessity", etc. In a sense, they are the universals of science.

1.2. SCIENCE AS KNOWLEDGE. A TYPOLOGY OF KNOWLEDGE

The term "knowledge" is traditionally used in the following three senses. The first sense refers to a predisposition, ability, capacity, skill, etc., based on "knowhow"—how to do or make something. The second sense is implied when knowledge is identified in general with any cognitively significant information (in a particular case, with information corresponding to reality). The third sense is linked with the interpretation of knowledge as a special cognitive unit (an epistemological taxon). In this sense, knowledge is interpreted either as everyday or as scientific.

The first sense of the term "knowledge" (knowledge as practical skill, or handicraft) is considered in 3.3. Here, we will be concerned with the epistemological nature of knowledge in senses two and three.

The typology of knowledge in sense two may be based on the character of the cognitive relation of the subject to the truth, the forms of its fixation, verification and recognition.

The study of the nature of the phenomenon of recognising the truth was begun in the epistemological inquiries of Immanuel Kant. "The holding of a thing to be true," he wrote, "is a phenomenon in our understanding which may rest on objective grounds, but requires, also, subjective causes in the mind of the person judging" (151, 496). An objectively sufficient ground for recognising something as true is, according to Kant, a judgment that is "valid for every rational being". A judgment that has its ground "in the particular character of the subject" is subjectively sufficient (ibid.). The objective validity of judgments is determined, in Kant's view, by the object, which also constitutes the substantive aspect of knowledge—the truth abstracted from subjective evaluations and recognition. The subjective validity of judgments, or the recognition of their truth on subjective grounds, has three stages: opinion, belief, and knowledge.

"Opinion is a consciously insufficient judgment, subjectively as well as objectively. Belief is subjectively sufficient, but is recognized as being objectively insufficient. Knowledge is both subjectively and objectively sufficient" (151, 498).

Leaving aside the Kantian approach, let us consider what positive meaning may be discerned in the "sufficiency" of certain grounds for recognising something as true. Let us bear in mind that in scientific knowledge the truth is given to the subject in accordance with the principle of sufficient reason.

The principle of sufficient reason as formulated by Leibnitz reads: no scientific proposition "can prove to be correct without a sufficient reason why that is so and not otherwise" (160, 41). A more detailed formulation of this principle is to be found in Theodor Lipps. According to the latter, this principle imposes three requirements on any scientific proposition: (1) in any proposition, the subject must contain a sufficient ground for the predicate; (2) propositions must have sufficient grounds in experience and in the laws of thought; (3) propositions must have grounds in other propositions (i.e., they must be deducible from general propositions) (164, 149).

The principle of sufficient reason delimitates between scientific and non-scientific knowledge. The existence or otherwise of sufficient grounds for asserting the truth of a certain proposition in a definite cognitive context is the qualitative basis for evaluating its nature.

From this standpoint, all non-scientific knowledge has no necessary basis. The basis of scientific knowledge is apodictic. The apodixis of knowledge, which means that this knowledge is compulsory for any object of cognition, is constituted by the internal structure of knowledge, its logical organisation interpreted as structural ordering.

Thus the conditions of the truth of scientific knowledge established on the basis of the principle of sufficient reason are discursive and rationally verifiable. That is the reason why Plato, in defining the epistemological difference between knowledge and opinion, declared the conditions of the former to be rational, and of the latter, sensuous. Inasmuch as "opinion", in terms of the Kantian typology of forms of recognition of the truth, is a form of conscious recognition of the truth from the position of insufficient grounds, the subjective relation to the truth in the framework of opinion will be, as it were, sensuous (resulting from a kind of expectation, guessing, presentiments, etc.) rather than rational-discursive (resulting from proof).

The description of various modifications of cognition constituting different forms of recognition of the truth would be incomplete without a reference to "doubt" and "belief" which, on the analogy of "opinion", are included in the domain of the sensuous.

Doubt is perceived as a state of subjective uncertainty about the truth of a proposition, which is largely determined by the feeling of lack of agreement between this proposition and the subject's entire past experiences, a feeling that prevents the subject from seeing it as true. Doubt therefore signifies the absence in the subject of a unitary consistent view of a certain object, a simultaneous existence in the subject of contradictory opinions which, in the absence of sufficient grounds, does not permit him to form an unambiguous judgment of this object.

Like doubt, belief is also a form of sensuous recognition of the truth—the difference between them being that belief is an awareness of harmony rather than disagreement between a judgment and our past experiences, an awareness which, as opposed to doubt, expresses an element of conviction and indubitability of the truth of something for the subject.

To complete our analysis of this theme, let us characterise surmise. In this case, we have a form of the subject's unintentional fixation of the truth "in itself" rather than a conscious recognition of something as true. Of this nature is, say, the atomism of antiquity, which, unlike modern atomism, was not comprehensively substantiated.

The typology of knowledge according with the third sense of the term "knowledge" is established on various grounds, of which the following appear to us to be the most fundamental.

The cognitive basis. On this criterion, knowledge is classified into discursive and intuitive-imageful (emotional). Knowledge in the broad sense functions as an instrument of verifying the truth, as a complex of demonstrations conditioning the self-obviousness of the truth for the subject. However, taking into account the fact that the truth of different types of knowledge is verified in different ways, it is reasonable to single out different types of obviousness.

Apart from the intuitivist interpretations, obviousness may be perceived as:

(a) something psychologically obvious—obvious in the individual-personal sense (personality background, etc.), something that signifies the subject's confidence in the truth of a fact on the basis of his personal experience; this type of obviousness can be illustrated by the variation in the force of inductive proof containing a recurrent feature—for the non-specialist, the specialist, and the highly qualified specialist;

(b) something logically obvious; that is the obviousness of proof, or apodixis; this type of obviousness is always mediated, being the result of demonstration, substantiation, proof, etc., as illustrated by any mathematical theorem;

(c) something immediately obvious—a form of the fixation of the state of affairs on the surface, so to speak, on the basis of the very essence of the situation; for example, the proposition "white is not black" is self-obvious due to its "self-expressiveness".

It follows from this that discursive knowledge is obvious in a logical sense, while intuitive-imageful (emotional) knowledge is obvious psychologically. Being logically explicit, discursive knowledge is, ideally, apodictic.

As opposed to discursive knowledge, emotional knowledge is "instinctive" and personalised. It is directly crystallised in communication, in which the subject forms a conception of the essence of events from an ensemble of barely perceptible nuances. Thus a person mostly "knows" when he or she is trusted (or not trusted), believed (or not believed), etc. Following from an allround evaluation of a situation actually experienced by the subject, this knowledge is not discursive; at any rate it is not as a rule rationalised or generalised to cover similar situations. For example, there could hardly be a single-valued solution to any attempt to generalise or rationalise the knowledge (of the predestination of man, community of human fates, insignificance of individual actions, etc.) that spontaneously emerged in the Pierre Bezoukhov—Davou situation in *War and Peace,* in which an "uncompromising conqueror" lets an arsonist and rebel escape with his life.

In this and similar cases we are dealing with knowledge/understanding based on image-symbolic forms which cannot be clearly expressed in clear discursive logical terms. In this sense, the minimal condition of the possibility of scientific activity, intended to produce scientific discursive knowledge as its output, is the possibility of conducting research in a natural language which is, apparently, the lowest threshold of scientific rationality.

Scientific experience is the sphere of existence of discursive knowledge, while emotional knowledge exists in non-scientific (everyday, moral, artistic, etc.) experience.

The social basis. On this criterion, knowledge is classified into personalised knowledge, problem knowledge, and subject-matter knowledge, reflecting the dynamics of sign translation of knowledge in a socium. All of man's activity involves signs. A widespread conception of man associates his *differentia specifica* with the ability for acting as a sign-symbolic being operating with symbols and signs. Complex sign activity, as close study has shown, developed because the mechanisms of biological coding and translation of information were too limited and inadequate for the self-preservation and progress of *Homo sapiens*. In the course of his evolution, man objectively comes up against the fact of natural limitedness and lack of universality of the means of biological information; it transpires that sociality, i.e., the special structure that arises in interindividual communication and interaction, is by its very nature non-biological and is not translated biologically. That is why other, non-biological means are required for the reproduction of sociality. In this way, culture arose as a mechanism of non-biological sign translation, as a social code ensuring the fixation, storage and transmission of human values in the broad sense, which makes them products of subsequent consumption.

What is the dynamics of the realisation of the social code?

At early stages, a personalised type of translation of knowledge exists (initiation rites for neophytes in primitive societies, myths as instructive narratives describing the deeds of ancestors, etc.); it is extremely imperfect, as the whole of the information gained at great cost may be lost through the frequently accidental loss of its carriers. To this type of translation of knowledge corresponds personalised knowledge (in the sense of *technē*, or individual ability) which is an individual's unique property.

Later, this type of translation of knowledge is replaced by the professional one—transmission of knowledge to members of a unified association of men formed on the basis of community of the individuals' social roles; here, the place of the individual is taken by a collective guardian, accumulator and translator of group knowledge/art, which is somewhat more perfect owing to the expansion of the social field of the carriers of knowledge—a kind of insurance against irretrievable loss of knowledge in case of loss of its individual carriers. To this type of translation of knowledge corresponds problem knowledge rigidly linked with the concrete cognitive tasks arising as man encounters a certain typological class of problem situations. Of this type are, for instance, certain archaic types of ancient Oriental knowledge constituting a set of prescriptions for subjects solving concrete tasks or problems. Problem knowledge, just as imperfect as personalised knowledge, was doomed, as it was inadequate to the growing needs of the cultural progress of mankind either cognitively or practically (in view of the devaluation and trivialisation of the available problem fund).

In its turn, this type of translation of knowledge is ousted by the universal conceptual code, the most perfect of the three; here the subject joins social activity as a citizen, without tribal, professional or any other similar limitations. To this type of translation of knowledge ideally corresponds subject-matter knowledge—the product of cognitive assimilation of a certain fragment of reality by the subject. Unlike problem knowledge, associated with the spontaneously empirical, pre-scientific stage in the development of the intellect, subject-matter knowledge, personifying science, is not a set of instructions for the cognising subject; in fact, it does not describe subjective activity at all. Summing up the cognitive reflection of a certain domain, it describes something objectively existing.

The process of science formation—the transition from personalised knowledge and problem knowledge to subject-matter knowledge—appears to involve the following elements.

(1) Systematisation of particular solutions and methods for dealing with problems, their fixation in some integral form permits, due to association with corresponding typical conditions, an abstraction from the unique situation giving rise to these problems.

(2) Factoring out the particular conditions does not merely signify the study of problems in general form—it assumes, as it were, an unconditional description of the entire domain generating the corresponding types of problems, which is, in fact, the starting point of science.

To take an example. Knowledge about forests, originally accumulated by the individual peasants who exploited them, functioned as personalised knowledge (situational knowledge) and was handed down from father to son. The requirements of industry (capital construction, shipbuilding, etc.) eventually led to the need for a systematic description and knowledge of forests: what areas were dominated by what kinds of trees, and why; what methods and sites of logging were the most profitable, and so on. This led to generalisation of spontaneous-empirical knowledge in this field. As a result, gradually evolving scientific dendrology came to cover allround rather than fragmentary information about forests as a natural geographical phenomenon (G. F. Morozov, V. V. Dokuchayev, N. A. Mikhailov). The main point in the transformation of forestry into dendrology was the identification of the domain of knowledge, the fixation of the information about the character of the reality under study in general form, the presentation of forest as an idealised object and a topic of special consideration; consequently, the main point was the transference of knowledge from a utilitarian applied sphere to the theoretical one. That is, in general outline, the dialectical process of the development of personalised knowledge (problemocentrism) into universal knowledge (with the focus on subject-matter), of non-science into science.

The subject-matter basis. On this criterion, knowledge is classified into natural, social, humane, and technical.[2]

In terms of *reflection of the essence,* knowledge is divided into phenomenalist and essentialist. Phenomenalist knowledge embraces qualitative theories mostly performing descriptive functions (many branches of biology, geography, geology, etc.). In contradistinction to this, essentialist knowledge is explanatory, resorting mostly to quantitative methods of analysis in the study of various domains. Naturally, phenomenalist theories are no substitutes for essentialist ones, just as the descriptive function of theory cannot supplant the explanatory function.

The development of scientific cognition leads, sooner or later, to the transformation of phenomenalist theories into essentialist ones—through perception of the essence, fixation of the causes of the phenomena under study, etc.

In relation to the forms of thinking and modes of categorisation, knowledge is classified into empirical and theoretical. The typology based on the empirical vs theoretical dichotomy orders knowledge from the standpoint of the theory of thought forms, taking into account their functional role in the structure of intellectual activity; this opens up additional possibilities for determining the qualitative nature and the specific features of available knowledge.

Knowledge corresponding to the empirical level is largely connected with generalisation of factual data, experimental dependences, regularities, inductive laws, etc. Knowledge of the theoretical level is more abstract, emerging as a result of immanent development of theoretical problem areas.

An essential difference between empirical and theoretical knowledge lies in the employment of different thought forms. Knowledge connected with the empirical level is formed as a result of sensuous fixation, or recording. Knowledge connected with the theoretical level is formed as a result of semantic interpretation, conceptualisation, and rationalisation.

The dialectical interrelation between empirical and theoretical knowledge sooner or later results in the transformation of the former into the latter, through the necessary substantiation of the former. Thus the Kepler laws, formulated by the author as inductive generalisations, were deduced, in the course of the development of classical mechanics, as theoretical knowledge from the more fundamental Newtonian law of gravitation.

In terms of *functional purpose,* knowledge is classified into descriptive and explanatory,[3] fundamental and applied.

The term "fundamental science" is applied to research reflecting the most general, profound, and essential aspects of reality (12; 37). When the motives, goals and tasks of such studies are discussed, they are often described as "pure" or "basic" (39; 123; 172, 33). Although these nuances reflect real elements of the methodological analysis of scientific knowledge, we shall use the terms "fundamental" and "pure" as synonyms, as is fairly frequently done in epistemological research.

Contrasted with fundamental or pure science are applied sciences connected more or less directly with practice and directed towards the solution of economic and production tasks, towards increasing our power over nature. Some researchers (10), pointing out the conventional nature of the definition of applied sciences, believe that fundamental knowledge can become applied if it is used in or applied to the system of another domain of fundamental knowledge. The fundamental-applied relationship is thus linked with the direction of the movement of knowledge in the interacting sciences. These conclusions stem, in particular, from the nature of knowledge and methods in a number of applied natural sciences, such as applied physical optics, applied infrared spectroscopy, etc. In these applied sciences, knowledge and methods are oriented both towards their use in other sciences, in laboratory work, and towards their employment in the sphere of material production. The situation is in fact the same in some branches of applied mathematics, whose methods and content also warrant the same conclusions.

In our view, science and knowledge become applied in character only in the solution of practical tasks. Application of group theory to pure crystallography does not make the former an applied science, and neither does the application of the apparatus of differential and integral calculus to theoretical mechanics make the latter applied mathematics. For this reason, of prime significance for establishing the specificity of applied studies is analysis of practice.

Without rejecting the importance for cognition of other forms of practice, we must stress the special significance for the study of applied sciences of such forms as industry and experiment. The links of applied sciences with industry, with the tasks of material production, have been well studied and are, one might say, fairly obvious. The effects of the requirements of production and of scientific experiments on the emergence of applied mathematics or, at least, on the applied aspects of mathematical knowledge have been studied much less. The emergence and development of applied knowledge of man has been almost entirely left without methodological interpretation in the context of the solution of production tasks. This applies to knowledge of man both in its scientific form (with the exception of psychology) and in the artistic one.

The dynamics of the fundamental and applied aspects of scientific knowledge is not exhausted by the movement in one direction, so to speak—by the generation of applied sciences in the process of solution of production, industrial tasks. The history of science provides numerous examples of transformation of applied sciences into fundamental ones. These transformations point, in particular, to the inadequacy of the differentiation between fundamental and applied knowledge on the basis of motives and goals. In itself, knowledge does not contain goals and, still less, motives which might be distinctive features here. The goals and the motives belong to the human subject, lying outside the cognitive structures of knowledge.

A more effective criterion or basis for the division of sciences into fundamental ("pure") and applied is the depth and generality of laws, models and practical schemata in cognition, or, putting it differently, the degree of concreteness of the laws, models, and practical schemata. The continuity of the development of scientific knowledge conditions, at each given moment, the existence of the tendency towards interpenetration of fundamental and applied studies. For this reason, it is rather difficult to draw a boundary between fundamental and applied knowledge at each concrete stage in the development of science. Therefore factors that may be called extrascientific often act as the ultimate argument in the division of science into fundamental and applied. Thus applied sciences received the status of separate branches in the curriculum of the Ecole Polytechnique in Paris founded in 1795 (11, 56)—the first occasion on which fundamental and applied sciences were taught as different subjects. In this case, the division was based on considerations of teaching methods, along with arguments bearing on the content of these disciplines.

The concepts of fundamental and applied knowledge are important characteristics reflecting the development of scientific knowledge and essential forms of differentiation and integration in science. Applied studies are not just applications of one science in the sphere of another. Synthesis of sciences has occurred on many occasions, but it has not always resulted in the emergence of applied sciences. New knowledge resulting from synthesis or integration of two scientific disciplines is not enough for applied research. Applied science is characterised not only by the emergence of new knowledge in the process of application of a fundamental science but also by the solution of industrial tasks, the tasks of production. The nature of applied mathematics is largely debatable because the content of its non-mathematical goals is not always the interests of material production. Mathematical tasks arising outside mathematics and solved in applied mathematics (9, 29) do not always come from industry. Mathematical instruments can be used in the design of planes and rockets, but they can also be employed in the modelling of the Peloponnesian War, of the rise and fall of world empires, or of the analysis of the style of ancient Russian authors. This kind of heterogeneity of the tasks, which we do not, as a rule, find in the applications of chemistry or physics, significantly affects the status of applied mathematical studies.

In terms of the *quality of the information contained,* knowledge is classified into probable and reliable, analytical and synthetic. "Probability" and "reliability" are modal characteristics of knowledge expressing the degree of its substantiation. Knowledge is considered to be reliable if there are grounds to assert that its truth has been established. Knowledge is considered to be probable if there are no firm grounds for confidence in its truth and it needs further logical or practical substantiation. The dialectics of the development of knowledge is subject to the law of transformation of probable knowledge into reliable knowledge through revealing the foundations for its truth.

The terms "analytical" and "synthetic" describe knowledge from the standpoint of the non-triviality of establishing its truth. Analytical knowledge is a set of analytical assertions whose truth is immediately obvious and depends only on the meaning of the terms contained in them, requiring no further explication. Thus the proposition "any daughter had a mother" is analytical. Synthetic knowledge is a set of synthetic statements—statements whose truth cannot be established directly, requiring as a rule a non-trivial factual procedure. Thus the statement "any body is in a state of rest or rectilinear and uniform movement if the resultant of the forces applied to the body equals zero" is synthetic.

The division of knowledge into analytical and synthetic is relative and has no meaning outside the framework of a certain fixed semantic system.

In terms *of cognitive genesis,* knowledge is classified into a priori knowledge and a posteriori knowledge. To avoid any misunderstandings, let us stress that there is no a priori knowledge as such, all knowledge is a posteriori. While rejecting the justifiability of using the terms "a priori" and "a posteriori" in some absolute sense, we are convinced that it is justified to use them in a relative sense with a view to the "functional-operative" role and purpose of definite types of knowledge in the cognitive process. On this approach, "a priori knowledge" is knowledge as premise or basis ensuring the actual unfolding of the cognitive acts in which derivative and, in this sense, a posteriori knowledge is obtained.

Thus the a priori vs a posteriori dichotomy (in the sense of premise vs derivative knowledge) is justified by referring to the complex composition of scientific consciousness consisting of (a) the layer of structures formed in the past (the available stock of knowledge) and (b) the layer of structures actually formed in the present (frontline science). The epistemological role of ingredient (a) is especially significant from the standpoint of the functioning and development of knowledge. Omitting the details, the basic functions of this ingredient are as follows.

(1) Inasmuch as any knowledge is built on certain premises as its foundation (as first shown in general form by Friedrich Adolph Trendelenburg in his evaluation of Hegel's project for a "premiseless" logic), premise knowledge comprises the conditions for the functioning and progress of the cognitive sphere. In cognition, it is impossible to begin entirely *ab ovo*—cognition cannot be started from the start but only continued from available premises.

(2) A most important function of premise knowledge is meaningful determination of cognition. This determination is usually referred to as determination by the cultural background, by the style of thinking, etc. As V. S. Shvyryov correctly pointed out, certain limiting implicit cognitive premises are necessary in science—premises "that are the necessary framework for the scientific-theoretical consideration of the world" (113, 81).

In the broadest sense, these premises are standard programmes of research coinciding with the types of scientific rationality which are accepted as necessary in the appropriate cultural-historical contexts.

Yet another function of premise knowledge ensures the continuity of cognitive activity. In this function, premise knowledge is a kind of criterion of the intelligibility and acceptability of newly obtained knowledge. That is probably the reason why scientific knowledge that does not accord with premise knowledge is called "crazy".

In terms of *epistemological status,* knowledge is usually classified into everyday and scientific. The idea of this classification dates from a remote past. Already in antiquity, the dichotomy was suggested of scientific (theoretical-rational, logically ordered) vs Socratean (atheoretical, logically unsystematised) knowledge, or, as it later came to be called, the knowledge of the life world.

The epistemological status of these types of knowledge is determined by the essence of the social institutions in which it is produced. Scientific knowledge produced in science as a specialised branch of social production satisfies definite standards regulating the parameters of the final "output product". Everyday knowledge produced in the framework of spontaneous activity in the everyday "life world" is, naturally, epistemologically unstandardised. The differences between scientific and everyday knowledge is therefore formulated in terms of the character of the objects they describe, the mode of reflection, type of categorisation, etc.—in a word, in terms of the specificity of the cognitive activity itself and of its products in the two distinct cases, where that specificity is determined by the orientation towards the criterion of scientificity.

The sphere of everyday cognition is highly multiform, covering as it does common sense, beliefs, knowledge of signs, generalisations of everyday experience recorded in traditions, legends, edifications, and the like, intuitive convictions, presentiments, etc. Everyday knowledge is extremely stable. Being a generalisation of recurrent mass phenomena and processes, it constitutes the basis of the individual's practical attitude to life and the world in general (the choice of values, goals, etc.). Enabling the individual to organise in this way his everyday activity, everyday knowledge is of fundamental significance for man as a natural-social being. It is appropriate to stress this in view of the raising of scientific knowledge to an absolute, characteristic of scientism—the view of science as a universal instrument of solving human problems, as "the measure of things", of what is that it is, and of what is not that it is not (182, 173). While rejecting the anthropological critique of scientism by Husserl, Heidegger, Jaspers and others, we would like to stress that the reduction of human experience, so diverse in its manifestations, to scientific experience is obviously untenable, since it has never covered and expressed, and will never cover and express, all that is human in man.

The epistemological relations of everyday and scientific knowledge have a dynamism of their own. On the one hand everyday knowledge works as a set of premises or assumptions in relation to scientific knowledge. On the other hand, encroaching on the domain of everyday life, scientific cognition modifies everyday knowledge, reshaping it on a scientific basis.

A specific feature of everyday knowledge is its correspondence to the pre-reflective stage in the development of the intellect, free from the control and analysis of its own resources (the procedures for obtaining, organising, and transforming knowledge). Endeavouring to deduce the truth from sensuous reality, everyday knowledge belongs, according to Hegel, to the pre-rational, pre-theoretical sphere. The mechanism that translates everyday knowledge into scientific one is substantiation. Everyday knowledge records the truth, of course, but it does so in an unsystematic non-specialised manner, leaving its foundations unexplicated.

As Alexander Luria has shown, "from the social-psychological standpoint, the operations of logical deduction from premises have no universal significance at all" (53, 54). Initially, when cognitive processes unfold on the spontaneously empirical plane of directly observable action rather than on the verbal-logical plane, individual, personality-related complexes of convictions derived from everyday experiences by incomplete induction prevail over intersubjective universally valid complexes of logical proof (for the latter are not yet formed as such). At this stage, there is no trust yet in logical premises as an ingredient of a compulsory "system of verbal-logical relations..., and the operations of logical deduction from premises do not yet have the same significance for obtaining new knowledge which they acquire... when theoretical forms of activity develop and become widespread" (ibid., 56).

Later, as "proto-logical", pre-scientific, practical everyday thinking functioning on the principle of non-reflective perception of connections in concrete situations is overcome, and rational-theoretical relations to reality develop, available knowledge is ordered and logically systematised.

Chapter 2

THE GENESIS OF SCIENCE

The emergence of science is an old problem with rich historical traditions. It has been and still is one of the pivotal problems of methodological, epistemological and historiographical studies. In fact, not a single fundamental study of these problems can ignore this issue—implicitly or explicitly, all of them deal with it. It will therefore be appropriate to consider this problem here in general form, without going into detail.

2.1. THE PROBLEM OF THE BEGINNINGS OF SCIENCE

It is widely accepted that to know a thing means to understand "why it is". But do we know and do we understand how, why and whence science emerged?

To overcome the numerous difficulties in the way of the solution of these questions let us separate the external beginnings of science from the internal ones, that is to say, "the beginning of something new in relation to the previous level" from "the beginning of something that is going to change, something that will be" (71, 37). It would appear that the external beginning is the boundary at which pre-scientific consciousness crystallises into science, while the internal beginning is precisely the point of departure of the history of science as distinct from its pre-scientific history. In the light of this it is appropriate to clarify this point: to solve the problem of the external beginning of science is to find out the dynamics of its development from pre-scientific cognitive forms; while to solve the problem of the internal beginning of science is to indicate a certain point in historical space from which it would be justified to date the beginning of the development of science itself. Taken together, these two solutions will constitute the solution of the general problem of the beginning of science.

The most remote cognitive premises of science are associated with the intellectual leap between the 8th and the 6th centuries B. C. when the transition from myth to logos was completed and cognitive structures took shape in the Middle and Far East, as well as in classical Greece, with which we still operate. The decisive conditions of this leap, i.e., the factors that overcame the mythological attitude to the world and thus objectively facilitated the formation of the rudiments of structures that led to the subsequent efflorescence of science, were as follows.

(1) The rejection of the "werewolf logic" of the myth, which interfered with the formation of such fundamental principles of scientific ideology as universality, invariance, etc. It is a well-known fact that the relation to reality of peoples at the lowest stages of intellectual development is based on direct sensual perception, which excluded the possibility of forming a picture of a nomologically self-sufficient, internally cohesive, self-identical reality. For example, members of the Aranta tribe typologise the world in terms of the "I see" vs "I do not see" opposition which obviously makes it non-self-identical. In this type of consciousness, the non-self-identity is determined by a kind of duplication of the worlds, which follows from man's ability to perceive an object as existing in the "invisible" world, apart from the visible one.

In a similar way, mythological consciousness identifies an object with the image in which that object is given to man, transforming the object to suit the various ways of its perception and making it go through metamorphoses alien to it. Everything therefore merges with everything else to form a single whole, everything is transformed into everything else in the mind of the carrier of mythological consciousness (in the mind of a child, at the early stages of philogenesis). No boundary is here drawn between the real and the unreal, between the objective and the subjective, between the true and the imaginary—it all appears to the mind as disjoint, accidental and, one might say, only possible and real but in no way necessary. It follows that the rejection of the "werewolf logic" of the myth was a great revolution in thought which asserted the picture of "non-bifurcating", "invariant", etc., reality that is not subject to arbitrary transformations depending on the properties of the human psyche.

(2) The replacement of the spiritual-personal relation to reality by the objective-substantial one. The destruction of the mythological identity of man and reality led to the formation of an "object ideology". The essential point here is that in the framework of that ideology reality emerged as an object struc-

ture subject to objective consideration. The assertion of this ideology immediately called forth numerous constructions whose cognitive status closely approximated that of science. Of this nature were, for instance, the extensive theogonic systems which, unlike mythological systems, were mediatedly discursive (noematic) rather than directly sensuous (aesthematic). They already contained an ingredient of scientificity detectable, say, in the assumption of rational construction of the world "out of itself" and not out of the individual's perceptions.

(3) The development of a natural interpretation of events. By this we mean a qualitative shift in the cognitive process produced by the requirement to invoke nothing but the natural, substantive, organic, etc., foundations of objects. The essence of this shift can be illustrated by the transformation of the principles of the interpretation of events in the framework of theogony and cosmogony. Analysis of genetic factors in these two cases shows the dynamics of the semantic resources which is subject to the law of consistent replacement of all supernatural and unnatural elements by natural ones.

It is necessary to touch in this connection on such an important point as the formation of a cause-and-effect typologisation of the phenomena of reality. Mythological consciousness, based on direct projection of human emotions, passions and experiences onto reality, based, in fact, on animating the world, resorted to a cause-and-meaning typology. There could be no other typology, since any event, being spiritualised, was not perceived as natural but as symbolising something in relation to the perceiver; it was seen as a sign of something designated and concealed by it, something connected in some way with the subject of perception. That was the background for the formation of a kind of symbolic parallelism of emotions and events with its invariable concomitants— "werewolf logic" and identification of thought and reality. But the gradual rejection of this logic and this identification, as described above, destroyed the basis for the cause-and-meaning typology. Indeed, if reality is independent of subjective affects, if the subjective and the objective are not identical, the "basis" for the phenomena of reality should be sought for in reality itself, not in the subject. On the other hand, the idea of the self-sufficiency of reality entailed the question of the mechanism of its internal organisation, integrality and cohesion, which, given the transition to interpreting reality through natural factors, resulted in the evolution of the cause-and-effect typology of phenomena that is the foundation of scientific knowledge.

Thus the identification of the external boundary of science with the qualitative leap from myth to logos throws light on the most remote premises of the development of science from pre-scientific consciousness. These premises include:

(a) the elimination of the mythological logic of the absurd, which is a generalisation of the rules for categorising reality on the basis of the cause-and-meaning typology. Inasmuch as a certain event was evaluated in this typology as symbolising some other event ($A \equiv B$, where $\equiv$ is the sign of equivalence) and not as self-identical ($A=A$), a kind of logic of the absurd became established with obligatory polysemy. The abandonment of this destructive logic and the transition to what is now traditional logic, with its laws of identity, consistency and excluded middle, was clearly the minimal condition of the emergence of science;

(b) the evolution of modes of cognition which, relying on discursive, rational complexes and foundations, constitute elements of object thinking oriented towards obtaining knowledge about the objectively existing.

Let us now consider the problem of the internal beginning of science. How did science in the proper sense of the word emerge? When and where did that happen?

A meaningful evaluation of this problem comes up against what might be called the difficulty of epistemological circularity. To find out from what entity science might reasonably be derived, i.e., what must be taken for the point of departure marking the beginning of true science as distinct from pre-science, we must know what science is. In other words, the definiteness of the beginning, of the historical point of departure cannot be established regardless of its subsequent relegation to what follows from this point, i.e., to science. Only an intensive analysis of science can therefore find out what its beginnings can be connected with and derived from. On the other hand, the effectiveness of analysis intended to show what science is is clearly minimal in the absence of the basic concept of the historical point of departure of science. Otherwise, two equally negative tendencies are possible here. (1) Structures basically alien to science, at one time realised in history (the question to be asked here is, in the true history or in the pre-history of science?), are uncritically and for this reason, of necessity, incorrectly included in science, which makes the problem of the epistemological definiteness and self-definiteness of science practically insoluble. (2) Elements from the past of science are critically yet unjustifiably excluded from science;

this approach is unacceptable, as it implies a modernising of science and a raising to an absolute of some of its actual features which may be not typical at all of science as such.

We are thus faced with a vicious circle: to perceive where science begins we must know what it is; but to find out what science is without either archaising or modernising it is impossible unless we rely on a firm previously given concept of the beginning of science.

Can this difficulty be overcome? One mode of solving it is through establishing a correlative and continually corrected connection between a certain limiting and highly flexible epistemological conception of science and its possible historical beginning. The positing of such a connection, as shown by the study of similar problems, is the only way of solving this kind of difficulty. The whole point is therefore to link the question of the historical beginning of science with that of its essence. In its turn, the question of the essence of science must not be discussed without some reference to its historical beginning. The solution of this problem thus requires (a) the working out of an epistemological standard of science, i.e., the setting apart of science as an epistemologically uniform and integral structure linked with a definite type of rationality, a mode of spiritual production specified by a minimal set of universal features; (b) an analysis of cognitive forms realised in history intended to ascertain their correspondence to the previously elaborated standard of science; (c) identification of (a) and (b).

Having posited the general principle of overcoming the difficulty of epistemological circularity, we can proceed with our discussion of the problem in hand.

There is no consensus among specialists as to how, when and where science emerged. We shall not evaluate or even list all the views on this issue that have been expressed. We shall merely reject from the outset the view of Herbert Spenser, who regarded the mind of an adult savage to be the most convenient point of departure for science. This position is based on the identification of science with any knowledge concerning reality; inasmuch as the savage had some knowledge of reality, he was believed to be involved with science.

Linking the beginning of science with rudiments of cognitive activity at the early stages of anthropogenesis, and identifying the subject of science with the primitive savage is not, in our view, a very profound approach; at the very least, it is inconsistent. It would be much more consistent to associate the starting point of science with the "research behaviour" of Anthropoids and thus connect the beginning of science with the intellect of apes rather than savages. But the absurdity of this line of reasoning is obvious.

The basic defect of Spenser's position lies, we believe, in his rejection of the genetic approach and unjustifiably placing science outside culture and human history. To show the untenability of Spenser's approach to the solution of the problem of the beginning of science, let us stress the distinction between two aspects of the concept of knowledge. On one plane knowledge is a mode of existence of consciousness. Since consciousness is intentional, it is impossible without knowledge, for it functions on the principle of realisation of the knowledge included in it. In this sense the existence of knowledge is by no means a prerogative of science but an attribute of any conscious activity, including, of course, non-scientific activity as well. Without knowledge, it is impossible for instance, to practise handicrafts, hunting, agriculture, or any other "standard" human activity unfolding as a certain technology for the implementation or achievement of some purpose. Naturally, the primitive savage possessed some knowledge gained through generalisation of social experiences—but this has little or no bearing on science as such.

This aspect of the concept of knowledge should be distinguished from that plane on which knowledge is identical with scientific knowledge, on which it can be equated with science.

Was the primitive savage in any way involved with science? A rigorous approach to this question requires, as was pointed out above, a method for the identification of "primitive knowledge" with the epistemological standard of science. What is that standard? As a basis for such a standard, let us take a model suggested by I. Rozhansky (84), which we shall reproduce here in brief outline. In Rozhansky's view, which we share, the model having a minimum of extremely general (even "weak") characteristics specifying science should be as follows.

(1) Any science is knowledge. Much more importantly, however, this knowledge is a result of an activity aimed at obtaining it. The determining feature of science is thus the existence of a special type of activity undertaken with the goal of producing new knowledge. From the sociological viewpoint, this activity can only be ensured by the availability of leisure, or a supply of time that becomes available when a group of persons is freed from material production; this supply of time is spent on development of nonmaterial production.[1]

That means that science in the proper sense of the word could not have appeared before the division of labour into intellectual and manual. Goal-directed, rather than spontaneous and sporadic, activity aimed at obtaining knowledge requires, apart from a definite category of individuals—the subjects of knowledge (producers, storers, transmitters)—a material basis (instruments and devices), its own methods (the means of obtaining knowledge, control of and instruction in knowledge), and a means of recording the results obtained (writing). A society possessing none of these attributes has no science either.

(2) The practice of science must be mostly motivated by cognitive problems as such arising out of a natural, disinterested need to know, and not by applied utilitarian problems built into the context of direct practical activity, so to speak. "Knowledge for the sake of knowledge" consolidates science as a special type of production opposed to other types of material as well as spiritual production (art, religion, etc.).

(3) To be scientific, cognition must be rational, i.e., it must completely exclude mythological, magical and similar notions based on belief in the supernatural.

(4) An agglomeration of disjoint knowledge obtained and functioning as an ensemble of empirical algorithms for the solution of experimental tasks does not yet form science. Scientific knowledge can only be separated out on the basis of proof through justifiable necessary derivation from a theoretical-fundamental consideration of the subject in "pure form".

Such is the most general epistemological model covering the typical aspects of science. In the light of this model, it is superfluous to demonstrate the untenability of Spenser's view of the origin of science. That sporadic primitive knowledge which the savage man attained at great cost through inductive generalisation of his activity (through trial and error) is not scientific in any respect whatever.

Now, how does what we call science take shape, after all? Where and when does it happen? As we shall see later, this question cannot be answered within the framework of epistemology alone—it also requires the study of the entire system of the material and non-material culture of mankind functioning as an integral whole. As a further step in our inquiry we shall therefore analyse the familiar types of human cultures in order to establish their actual potential for being the "generative structures" of science.

2.2. SCIENCE IN THE ANCIENT ORIENT

As the study of ancient science is not a goal in itself in the present discourse (we are only concerned with the roots and the dynamics of the structure called science), our analysis is only intended to elucidate the real potential of ancient Oriental culture for generating science by clarifying the social and epistemological features of that culture's functioning. Our immediate question therefore is: was ancient Oriental culture capable of generating science? Naturally, our discussion will only be effective if we rely on the standard of science fixed above.

Comparison of knowledge current in the ancient Orient with that standard shows the following.

(1) It should be recognised that the Oriental civilisation most advanced at the time (before the 6th century B. C.) in agriculture, the crafts, the war arts, and commerce (Egypt, Mesopotamia, India, and China) had worked out a certain body of knowledge.

Floods and the need for quantitative evaluation of the areas covered by water stimulated the development of geometry; intensive trade, the practice of handicrafts, and construction work motivated the development of computation operations and counting; navigation and cultic practices promoted the science of the stars, etc. Oriental civilisation thus had at its disposal a body of knowledge which accumulated, was stored and transmitted from generation to generation, helping to optimally organise human activity. It was pointed out above, however, that the existence of a body of knowledge does not in itself constitute science. Science is determined by the goal-directed activity of producing new knowledge. Was there any such activity in the ancient East? This question has to be answered in the negative.

In the most precise sense, knowledge was here produced by popular inductive generalisation of direct practical experiences, circulating in the socium on the principle of hereditary professionalism through (a) transmission of knowledge within a family in the process of assimilation of older members' skills by children; (b) transmission of knowledge, described as coming from a divine patron of a given occupation in the framework of a professional union of people (guild, caste) in the process of its expansion. Thus knowledge changed in the ancient East spontaneously; no critical-reflective activity of evaluating the genesis of knowledge existed—knowledge was recognised on a passive basis, without proof, through forcible involvement of an individual in social activity in the professional framework; there was no motivation for falsifying or critically renovating available knowledge; knowledge functioned as a set of readymade prescriptions for activity, which followed from its utilitarian, practical-technological character.

(2) One of the features of ancient Oriental science was its non-fundamental character. Science, as was pointed out above, is not the activity of working out prescriptions, technological schemata and recommendations—it is a self-sufficient activity of analysing theoretical problems, or "cognition for cognition's sake". Ancient Oriental science was, however, primarily concerned with the solution of applied tasks. Even astronomy, which would appear to be an entirely non-practical occupation, functioned in ancient Babylon as an applied art in the service of cults (the times of sacrifices were linked with periodically recurring celestial phenomena—the phases of the moon, etc.) or of astrology (studying favourable and unfavourable omens for acts of current military and civic policy, and so on). It was different, say, in ancient Greece, where astronomy was not seen as a set of computation techniques but as a theoretical science about the structure of the universe as a whole.

(3) Ancient Oriental science was not rational in the full sense of the word. The reasons for that were largely determined by the character of the socio-political structure of the ancient Oriental states. Rigid stratification of society, absence of democracy or equality of all before unitary civil laws and similar factors resulted, as it were, in a natural hierarchy of men comprising, for instance, heaven-sent rulers, perfect or noble men (tribal aristocracy or state bureaucracy), and common tribesmen. In the countries of the Middle East, states took the form of either open despotism or hierocracy, which also meant absence and rejection of democratic institutions.

The antidemocratic spirit of social life could not but be reflected upon intellectual life, which was also antidemocratic. Preference was always given to public authority rather than rational arguments or intersubjective proof (as such, they could not even take form against such a social background); accordingly, it was not a free citizen defending the truth by invoking just causes for his actions who was invariably in the right but the hereditary aristocrat and the powers-that-be. The absence of premises for generally valid substantiation and proof of knowledge (a situation caused by the esoteric "professional" rules for involving individuals in social activity and the antidemocratic nature of the social structure), on the one hand,

and the mechanisms of accumulation and translation of knowledge adopted in ancient Oriental societies, on the other, ultimately led to a fetishistic attitude to knowledge. Indeed, the subjects of knowledge, or individuals representing and personifying learning by virtue of their social status, were priests relieved from participation in material production and educated enough to engage in purely intellectual activities. Knowledge, though empirical-practical in genesis, remained rationally unsubstantiated within the confines of esoteric priestly science, sanctified by the gods, becoming an object of worship and a mystery. In this way the absence of democracy and the priests' monopoly of science conditioned by that absence determined the irrational, dogmatic character of science in the ancient Orient, making science a sort of semi-mystic, sacral occupation, a set of religious rites.

(4) Solution of *ad hoc* problems, primary concern with calculations of a particular, non-theoretical nature, etc., made ancient Oriental science unsystematic. It has been pointed out that the achievements of ancient Oriental thought were considerable. Thus ancient mathematicians of Egypt and Babylon were able to solve problems involving "first-degree and quadratic equations, equality and similarity of triangles, arithmetic and geometric progression, determining the areas of triangles and quadrangles, the volume of parallelepipeds" (54, 13); they also knew the formulas for the volume of cylinder, cone, pyramid, truncated pyramid, etc. Babylonians made use of tables of multiplication, reciprocal quantities, squares, cubes, solutions of equations of the type $x^3 + x^2 = N$, and so on.

However, ancient Babylonian texts provide no proofs for substantiating the employment of any given procedures, the need to calculate the magnitudes in question in a certain particular way rather than any other.

The attention of ancient Oriental scholars was always focused on some particular practical task, and no bridges were built between that task and theoretical consideration of the subject in general form. As the search for finding practical instructions and recommendations for acting in situations of a given type assumed no quest for universal proofs, the reasons for employing the appropriate procedures and algorithms were professional secrets, which made science rather similar to magic ritual or conjuring.

Besides, in the absence of a closely argued general consideration of an object it was impossible to deduce the information about that object that might be needed—the information, say, about the properties of geometric figures. That was why Oriental scholars and scribes were compelled to use cumbersome tables (of coefficients, etc.) for guidance in the solution of concrete tasks of an unanalysed typical class.

Thus, if we accept that each of the features of the epistemological standard of science is necessary, and their totality is sufficient for the specification of science as an element of the social superstructure and a special type of rationality, it may be asserted that science thus perceived was nonexistent in the ancient Orient. Although we know very little of ancient Oriental culture, the basic incompatibility of the properties of science existing here with the features of the standard of science is beyond doubt. In other words, ancient Oriental culture and ancient Oriental consciousness did not yet work out modes of cognition based on discursive reasoning rather than on prescriptions, dogmas and prophecies, on democratic discussion of problems, on debate from the positions of rational grounds rather than from the positions of strength of social and theological prejudices and authorities, and on substantiation rather than revelation as guarantee of truth.

In the ancient Orient, the limited body of knowledge and experience did not exceed the "confines of the traditionally supplemented and only very slowly and little widened collection of recipes"; the hand and the head were not yet fully separated here from each other (59, 47, 554).

Taking all this into account, our final evaluation is as follows: the historical type of cognitive activity and knowledge which evolved in the ancient Orient corresponded to the prescientific stage in the development of the intellect; it was not yet scientific.

2.3. SCIENCE IN CLASSICAL ANTIQUITY

The view is widely current in scientific literature (a view that we share) that the true cradle of science was classical Greece, whose culture at the time of its efflorescence (6th-4th centuries B. C.) gave rise to science.

Let us consider the distinctive features of this period, stressing from the very outset that we do not restrict the study of classical culture to an analysis of the unfolding of the first research programmes that can be described as scientific. We consider it important to fix those social and epistemological structures which, arising in antiquity, determined the formation of science as such here.

Let us begin with the socio-political causes for the unprecedented upsurge in Greek culture in the 6th-4th centuries B. C. The struggle between the demos and the landed aristocracy ended in the reforms of Solon (Athens, 594 B. C.), which substantially limited the real power of the aristocracy. The significance of Solon's reforms lay in the destruction and abolition of all estates and in the declaration of the principles of political and legal equality of free citizens, as recorded in the constitution of Clisthenes promulgated in Athens in 509 B. C., after the overthrow of tyranny.

These events had the following effect on the social superstructure, and in particular on problems of cognition. First, the individual endowed with civil freedoms was not depersonalised here, as he was in the tyrannical institutions of power depriving everyone of their rights, as, e.g., in the ancient Orient. The democratic form of the social structure in Greece which, on the one hand, assumed the participation of each free citizen in political life (in popular assemblies, public debate, voting), and on the other, encouraged in a practical manner the free play of the citizen's talents and potential, eliminated the prerogatives of birth and, moreover, was not conducive to humility in the face of rulers and bureaucrats, especially as the latter were elected and held their offices in succession. That is why the core of axiological consciousness among Greeks was the concept of the individual's personal dignity rather than his birth or social position.

Second, the establishment of generally valid civil law determined the extremely difficult transition from the interpretation of the order of social life in terms of Themis (divine statute sent, as it were, from above owing to a predestined order of things) to its interpretation in terms of Nomos (a statute that has the status of a legal idea duly debated and adopted). This move signified a kind of secularisation of social life, its liberation from the power of religious and mystical notions.

Third, the attitude to social law as a democratic norm whose civil excellence was proved in popular debate and accepted by the majority, and not as a blind force dictated from above, was based on the practice of rhetoric—the art of persuasion and argument. Indeed, as the strength of argument and critique became the instrument of making laws, the power of words grew, and the skill of handling words became a "form of political and intellectual activity, ... a means of conscious choice of a political line, and a mode of achieving justice" (41, 20). The Greeks even introduced a special deity in their

Pantheon—Petho, the embodiment of the art of persuasion.

Fourth, the citizens' legal equality and their subordination to unitary laws, as well as respect for the art of persuasion, had as their consequence a relativisation of human judgement. Since everything in the sphere of the intellect had to be substantiated, and any substantiation, subjected to criticism, could be substantiated in another and more sophisticated manner, each Greek had a right to his own opinion. That right was only infringed upon when private opinions came in conflict with effective laws. In other words, the universal principle of critique and search for a better substantiation was only invalid in situations covered by precise laws which, once adopted, could no longer be criticised.

We can thus state that basically the Greeks saw the truth as the product of rational proof based on substantiation and understanding and not as a product of dogmatic faith relying on authority. Greeks were not, of course, one hundred per cent rationalists (are there any such?); there were factors that restricted the Greeks' *ratio:* belief in destiny, in chance (*tichẽ*), which could not be controlled, influenced, or withstood, and so on. It should be pointed out, however, that these concessions to the supernatural were mostly made in the Greeks' everyday civil life, but not in cognition. As far as cognition was concerned, the Greeks drew a hard and fast line between the rational and the irrational, decisively excluding the latter from consideration. Thus, excluding the mythological concepts of the structure of the universe suggested by Hesiod, Pherecydes of Leros, Epimenides and others from the context of physics, Aristotle focused on the analysis of the pre-Socratics' "physiological" views of the universe.

We see that a most important result of the democratisation of the socio-political sphere in classical Greece was the development of the apparatus of rational logical demonstration, which later outgrew the boundaries of direct realisation of political activity, becoming a universal algorithm for the production of knowledge as a whole and as an instrument for translating knowledge from the individual to society. Against this background, science as demonstrable knowledge based on certain grounds could already grow; this is readily proved by the available historical data. For instance, the natural-philosophical "physiological" constructs of the pre-Socratics are qualitatively different from the conceptually related ancient Oriental and earlier Greek mythological constructions precisely in being based on logical proof. Thus the invariably popular proposition concerning the unity and at the same time the non-identity of all things appears in the "physiologies" of the pre-Socratics as an element of rational deduction rather than an element of a poeticised worldview characteristic of the ancient Oriental and Orphic myths.

If we take rational substantiation as the minimal necessary premise of science, i.e., cognition in the form of proof through appeal to actually verifiable (not mystical) reasons and grounds, this principle underlies (even if we discount the "physiological" natural science of the pre-Socratics, the ethics of Socrates, the astronomy of Eudoxus and Calippus) the planimetry of Hipparchus of Chios, Hippocrates' medicine, Herodotus' history, Euclid's geometry, etc. All these unquestionably fall within the domain of science.

To fix more exactly the premises for the emergence of science we must discuss the employment of slave labour as a feature of Greek life. At the level of social consciousness, the universal employment of slave labour and the fact that free citizens were relieved from participation in the sphere of material production were the reasons why the Greek had such profound contempt for anything that had to do with instrumental-practical activity, which was naturally complemented by a contemplative ideology, an abstract, speculative and artistic attitude to reality. The Greeks drew a line between the mind's free play with an intellectual object and the productive labour activity with a material object. The former was regarded as worthy of a free citizen and termed science, the latter was suitable to a slave and was called handicraft. Even such a highly artistic activity as sculpture had the status of a handicraft in ancient Greece, being connected with "matter". The outstanding Greek sculptors—Phidias, Polykleitos, Praxiteles and others—were therefore regarded, in fact, as just so many craftsmen. Art and handicraft were identified, both being covered by one word and concept—*technē*.

Interestingly, in science itself the Greeks separated true science from its applications, and interest in the latter was not approved of. Thus the Greeks opposed physics as the science studying nature to mechanics—an applied branch, the art of building technical apparatus, of designing and building machines. It is clear in this context why Plato reproached Eudoxus and Archytas for their studies in mechanics; Aristotle also disapproved of enthusiasm for mechanics. In mathematics, the art of carrying out concrete calculations fell in the domain of the lowly *technē*, while arithmetic was regarded as a respectable theory of abstract properties of numbers.

In what connection do we speak of contemplativeness in our analysis of the premises for the emergence of science? The point is that a necessary condition for the development of science is the use of idealisations, which cannot appear within a material-practical attitude to reality. Generalisation of the principles of instrumental labour activity involving various objects merely gives rise to abstraction—a fairly "standard" epistemological operation of identifying actually existing features, an operation that the higher animals can also perform. Abstraction, however, is obviously incapable of producing idealisation—the identification of features that do not exist in reality and cannot therefore manifest themselves in instrumental-practical impact on reality. Therefore, for idealisations to emerge, it is necessary to give up the material-practical attitude to reality and to accept the positions of contemplation; this end was achieved in ancient Greece.

The idealisations which figure in ancient Greek texts, and the associated purely theoretical questions, the special apparatus of intersubjective substantiation employed for the organisation of systems of knowledge, etc., were not, obviously, inductive generalisations of production activities. Whether we consider the propositions of Hipparchus' planimetry or the postulates of Euclid's geometry, the aporias of the Eleatics or the problems of *archē* which were of such interest to all pre-Socratics, the Pythagorean question of incommensurability or the search of Diogenes for the essence of man—none of this has any traceable links with material production. Generalisation of the practice of land surveying does not yield the concepts of the Euclidian straight line, plane, point, etc. Generalisation of the practice of a metal worker or potter will not result in Heraclitus' concept of fire as the first element of the universe, and so on. While giving rise to abstraction, practice blocks the emergence of idealisations as its logical continuation. No practical worker as such will concern himself with problems of the essence of the world, cognition, truth, man, the beautiful, etc. All these are basically "non-practical" questions, remote both from the sphere of mass production and from the sphere of the producers' consciousness.

Now, how did it become possible to formulate and discuss these questions? What were the reasons that made idealisations the core of the cognitive and cultural processes that gave birth to science? To some extent, these questions were answered in the above, where we stressed that the formation of ideal objects constituting the necessary foundation of science was based on contemplation, on orientation towards abstract-theoretical consideration of objects in their pure form, which prevailed in Greece. To this should be added that idealisation as a form of thinking was practically non-existent in the traditional societies of the ancient Orient. This factor must not be exaggerated, of course; abstraction was naturally inherent in the thinking of members of the ancient Oriental culture, and so was logical argumentation, otherwise there would be no point in speaking of ratiocination at all in that period. It is quite obvious at the same time that both abstraction and logical argumentation were extremely undeveloped in the Orient and thus incapable of forming a basis for the development of theoretical cognition, or science.

It would not be appropriate to discuss in this study the extremely complex question of the degree of scientificity of, let us say, the Greeks' natural-scientific doctrines compared to their ancient Oriental analogues on the meaningful plane. An evaluation of this knowledge in terms of form will be more definite and fruitful. Certain propositions are more or less obvious. It seems clear, for instance, that the ancient Greeks' cognitive potential for producing science was much more preferable than the corresponding potential of ancient Oriental culture. What we mean here is this. Although both in the ancient Orient and in classical Greece there was knowledge which could hardly be described as scientific in terms of meaning, only in Greece, and not in the traditional Oriental societies, did such forms of cognitive activity arise as systematic proof, rational substantiation, logical deduction, and idealisation—forms out of which science later developed.

The reasons for that lay in the specific features of the socio-political order in Greek society. We refer here to the institution of slave-owning democracy, which was favourable both to the development of an apparatus of intersubjective systematic rational-logical proof and to the elaboration of various devices for the designing of, and operation with, ideal objects.

On the basis of the above, the formation of science in ancient Greece may be reconstructed as follows. Mathematics in Greece did not differ at the beginning from ancient mathematics in the East. Arithmetic and geometry functioned as an ensemble of technical procedures in land surveying, falling in the domain of *technē*. Both in Greece and in the ancient Orient, these procedures were neither textually formulated in any detail nor rationally and logically substantiated. To become science, they had to be both adequately formulated and substantiated. When did that take place?

Historians of science have expressed various hypotheses on this score. It is assumed, for instance, that it was done by Thales in the 6th century B. C. According to a different view, this was done somewhat later by Democritus. There are other opinions. The actual facts, however, are not so important for us. It is important to stress that all this happened in Greece and not, say, in Egypt, where knowledge was translated from generation to generation verbally, and geometers were practical workers rather than theoretical scientists (the Greek word for them was αϱπεδουαπται, "those who tie a rope"). We can thus state that Thales and possibly Democritus played a role in giving mathematics the textual form of a theoretical-logical system. We cannot, of course, pass over in silence the Pythagoreans, who developed, on a textual basis, purely abstract mathematical concepts; or the Eleatics, who were the first to draw the line between the sensuous and the intelligible—a distinction previously unknown in mathematics. All this constituted the foundation of the formation of mathematics as a theoretical-rational science rather than an empirical-sensuous art.

The next element that is of the greatest importance for reconstructing the process of the emergence of mathematics is the working out of the theory of proof. Here, we should stress the role of Zeno, who contributed to the formation of the theory of proof by developing the apparatus of proof by the rule of contraries; and also of Aristotle, who carried out a global synthesis of all the known procedures of logical proof and generalised them as a regulative canon of research towards which all scientific, including mathematical, knowledge was oriented.

In this way the originally non-scientific empirical mathematical knowledge of ancient Greeks, in no way different from ancient Oriental knowledge, was transformed into science—through rationalisation, theoretical elaboration, logical systematisation and deductivisation.

Let us describe physics—the natural science of ancient Greece. The Greeks were familiar with numerous experimental data that later became the subject-matter of natural science. Thus the Greeks discovered the "attractive" property of rubbed amber and of loadstones, refraction in liquid media, etc. And yet experimental natural science never came into being in ancient Greece. Why was that so? The reasons lay in the specific features of the superstructure and social relations prevailing in antiquity. Taking the argument propounded above as our point of departure, we may say that the experimental type of cognition was alien to the Greeks because of (1) the complete dominion of the contemplative attitude at the time, (2) an aversion to separate "insignificant" concrete actions which were regarded as unworthy of the attention of the intellectuals—the free citizens of democratic polises—and unsuitable for the cognition of the world as a whole undivided into parts.

It is no accident that the Greek word "physics" is often used in quotes in modern studies in the history of science, for the Greeks' physics is something quite different from the modern discipline of that name. To the Greeks, physics was a "science of nature as a whole, but not in the sense of our natural science" (83, 9). The Greek word φυσις means "creation", so that the science of physics was a science of nature which included cognition through speculation on the origin and essence of the natural world as a whole and not through experimental testing. It was an essentially contemplative science very similar to later natural philosophy relying on the method of speculation.

The following two questions have to be answered: what are the premises for the emergence in antiquity of an ensemble of natural-scientific concepts, and what are the causes that determined their concrete epistemological character?

The premises for the emergence in antiquity of this ensemble of natural-scientific concepts include, first, the view, which asserted itself in the struggle against anthropomorphism (in the works of Xenophanes of Colophon and others), of nature as a naturally emerging structure (we hardly dare say "natural-historical structure") whose foundation is to be found in itself rather than in Themis or Nomos. The significance of eliminating anthropomorphism from cognition lies in the delimitation of the domain of the objectively necessary from the subjectively arbitrary. This provided organisational and epistemological grounds for the introduction of certain norms in cognition, for its orientation towards quite definite values, and in any case for preventing the merging in one whole of mirage and reliable fact, phantasm and result of rigorous research.

The second premise was the implanting of the idea of "ontological nonrelativity" of being, which followed "from the critique of the naive empirical worldview stressing continual change, of which a philosophical-theoretical version was worked out by Heraclitus.

The focus of Heraclitus' universe is the law of mutual transition, of continual self-restoration, conflict, and renovation of the substances of which the source and principles of motion he reduced to the mobile nature of fire, the first element of all that is.

The ideas of Heraclitus were sharply criticised by Parmenides; his treatise on *Nature* asserted that becoming was not and could not be the first principle of things. Parmenides stressed that the ideology of becoming, with its emphasis on the fluctuation of things, undermined the possibility of knowledge reflecting stable relations.

The views of Parmenides were shared by Plato, who drew a line between the world of knowledge correlated with the domain of invariant ideas, and the world of opinion correlated with sense perception recording the "natural stream" of that which is.

The results of this long-drawn-out controversy, in which practically all philosophers of antiquity took part, were summed up by Aristotle. His ideas were a further step in the development of this problematics: the object of science must be stable and general—properties that are absent in the sensually perceived objects; there was thus a need for a special subject separate from the sensually perceived objects (117).

The idea of an intelligible object that is not subject to continual change was epistemologically significant, as it laid the foundation for natural-scientific knowledge.

The third premise was the view of the world as a coherent whole comprising all that is and amenable to suprasensuous contemplation. This circumstance was of considerable epistemological significance for the prospects of the formation of science. In the first place, it helped to establish the fundamental scientific principle of causality; science is, in fact, based on setting down causal relations. Besides, in conditioning the abstract systematic character of the potential conceptualisations of the world, this view stimulated such an inalienable attribute of science as its theoretical character, i.e., logically substantiated reasoning relying on the conceptual-categorial apparatus.

These are, in the briefest outline, the premises for the emergence in antiquity of an ensemble of natural-scientific notions which were merely a prototype of future natural science, not yet science as such. The following are the causes of this state of affairs.

(1) An essential premise for the emergence of natural science in antiquity was, as we have pointed out, the struggle against anthropomorphism, which ended in the formulation of the programme of searching for *archē*, or the monistic basic of nature. Of course, that programme facilitated the assertion of the

concept of natural law. On the other hand, however, it also stood in the way of this concept—because of the lack of actual concreteness in that programme and the assumed equality of the numerous candidates for the role of the first element or *archē*. Here the principle of insufficient reason came into play; it prevented the unification of the well-known "fundamental" elements and the elaboration of the concept of a unitary generative principle (which might become law).

(2) Natural science did not exist in antiquity because it was impossible to apply the apparatus of mathematics in the framework of physics; according to Aristotle, physics and mathematics were different sciences pertaining to different objects and having no points of contact. Aristotle defined mathematics as the science of immobile being, and physics, as the science of mobile being. The former was fully rigorous, the latter, by definition, could lay no claim to rigorousness and therein lay their incompatibility (ibid). Unrelated to mathematics and thus devoid of quantitative methods of research, physics in fact functioned in antiquity as a contradictory fusion of two types of knowledge. One of these—theoretical natural science or natural philosophy was the science of the necessary, universal and essential in being which relied on the method of abstract speculation. The other type of knowledge—a naive empirical system of qualitative knowledge of being—was not even science, in the precise meaning of the term, for a science of accidental being given in sense perception could not exist, according to the epistemological views of antiquity. The impossibility of introducing precise quantitative formulations in the context of either of these types of knowledge naturally deprived them of definiteness and rigorousness, without which natural science could not take shape, as science.

(3) Some empirical studies were, of course, carried out in antiquity: cf. the establishment of the size of the Earth by Eratosthenes; the measurement of the visible sun's disk by Archimedes; the calculation of the distance between the Earth and the moon by Hipparchus, Posidonius, Ptolemy, etc. However, antiquity did not know experiment as "artificial perception of natural phenomena in which secondary and insignificant effects are eliminated and which has for its purpose the confirmation or rejection of certain theoretical hypotheses" (84, 15).

The reason for that lay in the absence of social sanctioning of free citizens' instrumental-material activity. Only impractical knowledge unconnected with labour was seen as seemly and socially significant. Moreover, true knowledge, that is, universal and apodictic knowledge, in no way depended on or came in contact with facts either epistemologically or socially. It is clear from the above that natural science as a factually (or experimentally) substantiated body of theories could not then take shape.

The Greeks' natural science was abstract and explanatory, it did not include an activity-oriented and creative component. There was no room here for experiment as a mode of affecting the object by artificial means with the aim of clarifying the content of the accepted abstract models of objects.

However, the formation of natural science as science requires more than the skill of ideal modelling of reality—it also calls for a technique of identifying the idealisations with the object domain. This could only be achieved under different social conditions and on the basis of socio-political, worldview, axiological, and other guidelines of cognitive activity substantially differing from those of ancient Greece.

In analysing the specific features of ancient science, we must thus stress that human thinking has developed from the outset "in many and divergent ways—among which one is the scientific" (196, 544). The reasons for that, as we have attempted to show, lay in the specificity of the peoples' entire socio-political and material-practical mode of the life, which, imposing an imprint on the character of their nonmaterial production, determined the possibility or impossibility of its functioning as the "generative structure" of science. Only that confluence of socio-cultural circumstances which occurred in ancient Greece was able to produce the necessary conditions for the emergence of science. It was here that such factors, necessary for the process of science formation, developed and took shape as intersubjectivity, universal validity, supraindividual character, substantionality, ideal modelling of reality, and so on. As they progressed and became consolidated, these factors ultimately conditioned that specific type of ideology and relation to reality which is called science.

2.4. SCIENCE IN THE MIDDLE AGES

The character of mediaeval science can only be understood if the entire system of mediaeval theological worldview is outlined, with its constitutive elements of universalism, symbolism, hierarchism, and teleologism. Let us consider these features.

Universalism. A specific feature of mediaeval thinking was gravitation towards universal knowledge, the desire to "grasp the world as a whole, to comprehend it as a kind of accomplished unity of all" (8, 2). The reason for that lay in the fact that the normative model of mediaeval knowledge was the classical epistemological model, described above, of true—universal and apodictic—knowledge, a model that was amply substantiated by the new socio-cultural and worldview materials. The actual basis of that model was the idea of the unity of man and cosmos, a unity rooted in their genetic community, or community in the act of creation; it followed from this that only he could know something who grasped the essence of divine creation; inasmuch as creation was universal, anyone having knowledge of it knew everything; conversely, he who had no such knowledge had no knowledge at all. Naturally, there was no place in this paradigm for partial, relative, incomplete or non-exhaustive knowledge; knowledge could only be universal—otherwise it was no knowledge at all.

Symbolism. Symbolism as an element of the mediaeval worldview was fully universal, covering both the ontological and the epistemological sphere. The sources of "ontological symbolism" become clear if one takes into account the radical nature of the propositions of creationism. Once created, any thing—from a mote of dust to nature as a whole—lost the status of ontological substantiatedness. Its existence, determined on a certain supreme plane, was not independent and was therefore necessarily symbolic, merely reproducing, embodying, or personifying the underlying fundamental essence of which it was an imperfect prototype and replica.

The ontological formula "The stamp of the Most High is imposed on all" produced as its epistemological equivalent the formula "Everything is filled with supreme meaning", which in its turn determined the conceptualisation of reality on the basis of a revived mythological and highly symbolic cause—and—meaning typology. The roots of the "epistemological symbolism" of the Middle Ages go back to the familiar precept of the New Testament: "In the beginning was the Word, and the Word was with God, and the Word was God" (189, 1, 1). The word here is an instrument of creation, an ontological element—but not only that. Passed on to man, it also figured as a universal way of comprehending creation, a means of joining in and reconstructing divine creative acts.

As concepts were directly identified with their objective analogues, and linguistic structures were universally hypostatised, the question of chimeras and fictions did not even arise; everything expressible in language, thinking, concepts and words was inherent in reality. The realistic isomorphism of concepts and real objects conditioned a sort of identity of the ontological and the epistemological, which figured as a condition of the possibility of knowledge.

In view of the genetically fundamental character of the concept in relation to reality, mastering and possession of the concept also signified possession of definitive knowledge about reality, derivative from the concept. Accordingly, the process of cognition of an object consisted in the study of the concept designating the object; this determined the purely bookish, textual nature of cognitive activity. And, since most available texts were sacred and, moreover, sanctified by divine authority, the ideal and the instrument of cognition was exegetics—the art of interpreting Holy Scriptures, that ultimate source of all possible knowledge.

Hierarchism. "All 'visible things' have the capacity for reproducing 'invisible things', for being their symbols. But this capacity varies from one visible thing to another. Each thing is a mirror, but some mirrors are smoother than others. This fact alone makes one think of the world as a hierarchy of symbols" (8, 34). Symbols were divided into "higher" and "lower"; membership in these two classes was determined by the closeness to or remoteness from God on the basis of the opposition of the celestial (intransient, noble) to the mundane (mortal, bestial). Thus water was "nobler" than earth, air "nobler" than water, etc.

Teleologism. Another attribute of the mediaeval worldview was teleologism, which consisted in the interpretation of the phenomena of reality as existing according to God's will in order to perform certain predestined roles. Thus water and earth serve plants, which, for that very reason, are nobler and occupy a higher rung in the hierarchy of values. Plants, in their turn, are food for animals.

Both logically and naturally teleologism culminated in anthropocentrism. The latter formed the basis for geocentrism. Mediaeval man was a highly ambivalent being. On the one hand, he was the summit of creation, the embodiment of the divine made in the image of his creator, and on the other, an object of the Devil's temptation, a vessel of sin. Man was always an object of struggle, the scene of conflict between the alternative world forces—God and the Devil. Man's real destiny was therefore the paramount question. This latter circumstance reinforced teleologism, of course. If we consider that God assumed human likeness and came down among men to suffer for humankind, and to show man his destiny, the world without man is certainly inconceivable—it would be meaningless. It was just as fundamental that the universal drama was played out on earth, where man dwelt. It was precisely the earth that was the stage on which the great drama was enacted in which the main protagonists were God, the Devil and man.

Evaluating these basic elements of the mediaeval worldview, we can draw certain epistemologically relevant conclusions.

Firstly, man's entire activity in the Middle Ages was channelled by religious notions. Nothing had the right to exist unless sanctified by the Church. Anything contradicting religion was interdicted by special decrees. This sort of orientation reinforced the element of contemplativeness, giving cognition a frankly mystical theological tone, which, far from encouraging advance in cognition, determined its regress or, at best, stagnation. Thus, the Middle Ages rejected the progressive theory of the genesis of nature suggested by the atomists of antiquity for the sole reason that the process itself of this genesis was seen as accidental (Democritus' απρονοησία) rather than as fatal and according with divine Providence. Another striking example was medicine, where all previously accumulated knowledge went by the board, and where mystical instruments like working wonders, prayer, relics, etc., took the place of proper medical ones (thus dissection, without which surgery was impossible, was anathematised as the greatest sin).

Secondly, the mediaeval picture of the world could not contain the concept of objective law, without which natural science could not evolve.

The mediaeval mind saw God as the cause of the interconnectedness and integrality of the elements of the world. The world was integral only insofar as there was a God that had created it. In itself, the world was devoid of cohesion; if God were to be eliminated, it would collapse, for all objects would lose their natural places in the hierarchy of things assigned by God. All objects being defined in relation to God rather than in relation to other natural objects, there was no place for the idea of objectness, the idea of objective universal interconnectedness and integrality, without which neither the concept of law nor, speaking more broadly, natural science could arise.

Thirdly, in view of the theological and textual nature of cognitive activity, intellectual effort was concentrated on the analysis of concepts and not of things, which were removed from consideration. Deduction was the universal method for establishing the subordination of concepts to one another, to parallel a definite hierarchical series of actual things. That which was logically deduced from another was already thought of as really subordinated to that other, as next in a series of objects of diminishing value, and this series was, in its turn, regarded as ontological. Since manipulation of concepts took the place of manipulation of real objects, there was no need for contact with the latter. Hence the basically a priori and extraexperiential style of speculative scholastic science doomed to fruitless theorising divorced from reality.

However, the view of the Middle Ages as mankind's intellectual cemetery would be a superficial one. Although mediaeval culture did not know science in the modern acceptation, specific branches of knowledge (we hardly dare to call them sciences) developed in it which paved the way for the evolution of science. We refer here to astrology, alchemy, iatrochemistry, natural magic, etc. It is remarkable, as far as our theme is concerned, that being a contradictory mixture of apriorism, speculation, and the most vulgar empiricism, these fields of knowledge little by little destroyed, in the process of their functioning and through their experiences, the ideology of contemplation, and achieved the transition to experimental science. The functioning of these disciplines, rightly considered (90) as an intermediate link between handicraft and natural philosophy, contained the embryo of future experimental science.

As we have pointed out, a necessary premise of science is the identification of objective regularly recurring situations in terms of experimental verification. In antiquity, this move was blocked by the contemplative attitude, which explains the impossibility of the evolution of empirically substantiated science in that epoch. The same attitude was an obstacle in the Middle Ages as well, only here, as distinct from antiquity, it had a purely religious, theological basis. An interesting point here is that the experiences of natural magic contradicted or, at any rate, did not accord with religious-mystical contemplation as a sort of ideological dominant. Indeed, religion is, in a general sense, an attempt to influence God's free will in a definite fashion (through cultic ritual) with the aim of achieving certain results (basically, religion is an appeal to certain "concealed parameters", an appeal reinforcing the determination of the behaviour of believers). Setting its hopes on God and being founded on faith, religion could not, naturally, provide any guarantees of the effectiveness of these attempts to influence divine will.

Like religion, natural magic was also an attempt to influence God with the aim of obtaining certain desirable results, but it set its hopes on empirical methods rather than on His free will. While religion was far from directing activity towards establishing empirically substantiated laws, natural magic necessarily assumed this kind of orientation; it differed from religion in its effective character, which was only ensured by experimental verification of abstract cognitive content. The latter brought magic closer to science and separated it from religion.

This aspect must not be exaggerated, of course. As V. Rabinovich precisely put it, "the mediaeval prescription as a special form of activity... is not just a set of instructions... but a form of activity in which actual performance is anticipated in verbal incantations" (80, 68). In other words, magic activity cannot yet be regarded as non-cultic. It was in fact accompanied by mystic spiritual rituals, sanctified by numerous prayers (verbal cult), etc., being a fully ecstatic and orgiastic craft. At the same time it could no longer be regarded as entirely cultic; in any case, it included epistemologically highly promising structures capable of transformation into experimental science—a possibility that was previously nonexistent. That was why "alchemy, which retained until the end close ties with magic, could evolve so smoothly into chemistry" (168, 16).

Our analysis permits the following conclusion. Mediaeval culture was a highly specific phenomenon in the history of European culture and of world thought. That specificity might be described in one word, as contradictoriness, i.e., as ambivalence and internal heterogeneity. On the one hand, the Middle Ages continued the traditions of antiquity, as illustrated by such cognitive phenomena as contemplative attitude, orientation towards inquiry into the general regardless of the particular, inclination towards abstract speculative theorising, rejection in principle of experimental cognition, recognising the primacy of the universal over the unique, of the stable over the evolving, of the supraindividual over the individual, and so on. On the other hand, the Middle Ages broke with the traditions of classical culture, preparing the transition to the completely different culture of the Renaissance. We find proof of this in the considerable progress in alchemy, astrology, iatrochemistry, natural magic, and other areas of knowledge having a purely experimental status. These elements integrated in a single whole, conditioned the contradictory character of mediaeval culture, which was perhaps of decisive importance for the destiny of science. The point is that, while retaining and reviving the skill of working with idealised constructs developed in classical natural philosophy, man's searching thought oriented itself precisely in that period towards achieving practical effects. And that was the decisive condition for the evolution of scientific theories of nature. We emphasise the word "condition", for natural science as science was not destined to take shape in the Middle Ages—for a variety of reasons.

(1) The idea of self-sufficiency of nature governed by objective laws was nonexistent in mediaeval culture; since nature was something created, it was controlled by the Maker's will. To change this paradigm, important ideological shifts were needed in the entire system of the worldview, which occurred only much later in the wake of Newton's and Voltaire's deism and Spinoza's pantheism.

(2) Another reason was the contemplative, theological and textual character of cognitive activity in the Middle Ages, which was so self-sufficient and deeply rooted in culture that it acted as a powerful worldview factor checking the progress of experimental science even in the days of Galileo.

(3) "Experimental" activity in science was semi-mystical in character, with the verbal element figuring fairly prominently—the adherents of natural magic believed in the mysterious force of verbal incantations. The concrete methods of natural magicians were not yet experiments in the generally accepted sense; they were rather magic rituals intended to summon spirits, otherworldly forces, and supernatural powers.

Strictly speaking, the mediaeval scientist did not operate with things but with the forces they concealed—with their ideal forms and proto-elements. Acts of experimental cognition unfolded therefore as ritual actions intended to establish contact with the next world; indeed, owing to the omni-present symbolism, the world of mediaeval man was two-dimensional, and the scholar functioned as a two-dimensional subject.

(4) The basis of the mediaeval picture of the world was qualitative ontology—Aristotle's theory of non-uniform and anisotropic space which asserted the "natural" dialectic of the elements and privileged status of different points and directions of motion in space. The epistemological positions were just as qualitative. We refer here to the traditional mediaeval doctrine of naive realism, which uncritically identified the subjective and the objective (cf. the formula *esse in intellectus—esse in re*), and ultimately interfered with adequate cognition.

The qualitative character of science, the separation between essence (*essentia*) and existence (*existentia*), object modelling, etc., made it impossible to evolve the concept of law, since the teleological concept of anthropomorphic causality (Aristotle's doctrine of four causes) blocked the development of the idea of reality dominated by natural, objective, and necessary connections.

We can thus state that mediaeval science was merely a stage in the development towards true science. We therefore believe it to be entirely wrong to place, as Duhem and Crombie do, the starting point of science (we refer to empirically substantiated science) in the mediaeval epoch, or to exaggerate the importance of the work of natural magicians, especially those of the Paris (Jean Buridan, Nicole Oresme, and others) and Oxford (Roger Bacon, Robert Grosseteste, etc.) schools (135; 131). Although some of the empirical results achieved by members of these schools anticipated subsequent attainments of classical science, these scholars cannot be seen as the founders of the creative method of science. As Lynn Thorndike correctly stressed, the science of Roger Bacon and others attached particular significance to working magic and did not go beyond the framework of fideist activity (187). The view of the Paris and Oxford schools as fountainheads of experimental science is an example of a non-historical approach to the analysis of the phenomenon of science, an unjustified trimming of factual data to suit an a priori research schema. True experimental science actually emerged in the Modern Times, and the point of departure here was the work of Galileo.

2.5. THE FOUNTAINHEAD OF CLASSICAL SCIENCE

The following processes accompanied the formation of natural science as science in Modern Times: the collapse of the archaic cosmosophy of antiquity and the Middle Ages under the onslaught of maturing naturalist ideology; the combination of the abstract theoretical (speculative natural-philosophical) tradition with the technical traditions of the handicrafts; the axiological reorientation of intellectual activity produced by the assertion of the hypothetico-deductive method of cognition.

The collapse of the cosmosophy of antiquity and the Middle Ages. Even a simple list of the causes of the intellectual revolution which brought down the classical and mediaeval view of the world and resulted in the formation of natural science as science would require a whole study, which would cover the production progress, the socio-political disintegration of feudal society; the Reformation, which eroded the solid structure of church ideology; Puritanism, which played a certain role in the evolution of rationalism; the consolidation of the institution of absolute monarchy; the strengthening of heliocentrism, which refuted the theological conceptualisation of reality in terms of the celestial vs. mundane opposition that retarded the development of cognition; the revival of the classical traditions of working with natural-philosophical idealisations; Protestant ethics with its idea of personal initiative, and many other factors. We shall therefore focus only on the principal items. The following notions and approaches were, in our view, the basis of the natural-scientific ideology which set the goal of obtaining knowledge about impersonal, blind, reproductive, self-determining everyday automatisms which emerge between mutually interacting objects.

Naturalism. Two circumstances helped to consolidate the idea of self-sufficient nature controlled by objective laws, devoid of any admixture of anthropomorphism and teleological symbolism, and conceptualised on the basis of the cause-and-effect typology, rather than the cause-and-meaning one.

The first circumstance was the development of such non-traditional theological doctrines as those of pantheism and deism. The dissolution of God in nature, which was at that time undoubtedly a form of atheism, made it difficult to worship the pantheistic god, on the one hand, and on the other, resulted in a sort of emancipation of nature, which was now equal in status to God and even prevailed over Him, given the concentration of cognitive interest on natural-scientific problems. Deism made a further step forward, actually asserting the possibility of natural objective laws, as it drew a line between creation as a supranatural act and the natural principles of existence of that which was created. The study of the former (the world's causes) was the demesne of metaphysics, while the investigation of the latter (the autonomously existing world as a consequence) fell within the realm of physics, with no points of contact between the two (the motto was: Physics, beware of metaphysics!).

The second circumstance was the development of medicine, physiology, anatomy, etc., which reinforced the idea of man's "animalism", his unity with organic and inorganic nature (man as a thing among a multitude of things), and destroyed anthropocentric teleologic illusions concerning man's privileged status in the world.

Combinatoriness. This is taken to mean a worldview approach to problems of the structure of reality which is opposed to the previously dominant symbolic-hierarchic approach. On this approach, any element of the world is a set of forms of varying degree of essentiality and universality rather than a qualitative whole integrally connected with other similar integralities in an all-embracing and all-permeating totality.[2] This was the basis for the view of the unity of the world as community of its forms—a view that undermined the qualitative perception of the world as an infinite multiformity. The entire diversity of reality was now described in terms of mechanical combinatorics of several fundamental forms responsible for certain qualities. Accordingly, to know reality meant to know the rules of combining the forms. The latter also determined such specific features of the new ideology as instrumentality and mechanicity, which played an important role in the formation of natural science as such.

Quantitativism. Combinatoriness formed the basis for the development of quantitativism—a universal method for quantitative comparison and evaluation of forms constituting any object: to know meant to measure. A considerable impetus to these advances in the methods for quantitative description of forms was given by the development of the apparatus of analytical geometry by Descartes and his followers, which substantiated the idea of the unity of geometrical forms and figures united by formal transformations.

It was also essential that qualities which had previously appeared incommensurable (thus Aristotle was unable to create the theory of value, although he came close to doing so) now proved commensurable, so that a picture of a unitary, homogeneous, and quantitative cosmos emerged in place of the hierarchised, heterogeneous, and qualitative one.

Cause-and-effect automatism. An essential contribution to the moulding of the image of the natural cause-and-effect cohesion of phenomena was made by Hobbes, who eliminated the last two of the four types of causes introduced by Aristotle—material, efficient, formal, and final. This worldview position, which was actively supported in scientific thinking (Galileo, Boyle, Newton, Huygens, and others) removed the shades of symbolism and teleology from the picture of reality and opened the way to its description in terms of objective necessity and regularity. We should also point out the increasing consolidation in that epoch of the monotheistic character of belief, which was absent in antiquity and which did much more than the classical ideas of obligation and order to assert the idea of uniformly and regularly determined reality.

Analyticity. "Among the Greeks," wrote Engels, "just because they were not yet advanced enough to dissect, analyse nature—nature is still viewed as a whole, in general. The universal connection of natural phenomena is not proved in regard to particulars; to the Greeks it is the result of direct contemplation" (59a, 46). In the Modern Times, a style of cognition quite different from that of antiquity asserted itself, cognitive activity was no longer in the form of abstract synthetic speculation but in that of a concrete analytical reconstruction of the plan, order and constitution of things, an ability for breaking them down into their basic constituents. The primacy of analytical activity over the synthetic one in the thinking of that period facilitated the formation of a system of physical causality which took final shape and was consolidated with the emergence of Newton's mechanics.

Geometrism. That feature of thinking, which we oppose to classical physicalism and mediaeval hierarchism, took shape as a consequence of the assertion of heliocentrism. The doctrine of heliocentrism was of immense ideological significance, since underlying the purely physical question raised by Copernicus, the simple scientific problem which he formulated, was something of extreme importance—the question of the position of man in the universe. Of revolutionary significance was, in the first place, the ontological aspect of heliocentrism. The ontology of antiquity and of the Middle Ages was based on Aristotle's doctrine of anisotropic and non-uniform space, which permitted the formulation of the idea of the five elements, including ether as the quintessence of being set against the conditions of mundane being and to formulate on this basis the antinomies of the heavenly and the mundane, etc., whereas Copernicus based his constructs on uniform and isotropic (Euclidian) space in which all points and directions of motion are of equal value. Inasmuch as Copernicus linked physical action in space with the points of concentration of the material substratum regardless of their position,[3] there were no qualitative ontological differences between heaven and earth in his theory. That meant the formation of the picture of a unitary cosmos; the development of this picture, which also involved questions of epistemology, permitted the substantiation of the doctrine of universal laws of nature.

We can now state the principal features of the new style of thinking which destroyed the archaic classical and mediaeval picture of the world and resulted in the formation of the object-naturalist conception of the cosmos that formed the premise of natural science as science. These are the features in question: the attitude to nature as a self-sufficient, natural, "automatic" object devoid of the anthropomorphic-symbolic element, given in direct activity and subject to practical assimilation; rejection of the principle of concreteness, of the naive qualitativist corporeal-physical thinking of antiquity and the Middle Ages; the formation of the principles of rigorous quantitative evaluation (in the evolution of commercialism, usury, statistics, etc., in the social field; and in connection with scientific advances in the invention and construction of measuring devices—clocks, scales, chronometers, barometers, thermometers, etc.); rigid determinist cause-and-effect typologisation of the phenomena of reality; elimination of teleological, organismic and animistic categories; introduction of causalism; instrumentalist interpretation of nature and its attributes—space, time, motion, causality, etc., which are mechanically combined together with ontologically basic forms constituting any object; the image of the geometricised homogeneous unitary reality controlled by unitary quantitative laws; recognition in dynamics of a universal method for the description of the behaviour of surrounding phenomena (formal geometrical schemata and equations in place of object models).

Combining the abstract theoretical (speculative natural-philosophical) tradition with that of the handicrafts. Science is constituted by the unity of empirical and theoretical activity. But in antiquity and the Middle Ages these two types of activity were epistemologically and socially separated from and placed in opposition to each other. Theoretical activity centred on seven classical liberal arts: astronomy, dialectics, rhetoric, arithmetic, geometry, medicine, music—and was not concerned with anything else. Empirical activity fell within the province of the mechanical, non-liberal arts, or the handicrafts. Some of the effects of this division bordered on the ridiculous. Thus, theoretical study of medicine was regarded as a scholarly pursuit and confined to the study of books, while the practice of medicine, that is, healing the sick, was considered to be unscholarly—merely the medical craft.

The situation in which theoretical studies were the province of abstract intellect, and empirical (experimental) research, that of concrete craft, seriously hampered the synthesis of empirical and theoretical levels and consequently made the formation of science impossible. Armchair scientists who ignored experimentation for psychological reasons (the nonprestigious character of experimental work), doomed themselves to fruitless system building and scholastic theorising. As for the craftsmen of the gilds, who did not study theory for social reasons (the barriers between the estates), they could not break out of the confines of creeping empiricism. We owe the disruption of this vicious circle and the radical change in the situation which led to a synthesis of empirical and theoretical activity and thus to the formation of science, to the social-practical processes which formed the core of the social life of that time.

As Zilsel correctly points out, science emerged when the "barrier between the two components of the scientific method broke down, and the methods of the superior craftsman" (i.e., empirical activity) "were adopted by academically trained scholars" (i.e., those trained in theoretical activity) (196, 555). That happened during the Renaissance as a result of the rapid progress of industry stimulated by the development of capitalism. As Engels wrote, "If, after the dark night of the Middle Ages was over, the sciences suddenly arose anew with undreamt-of force, developing at a miraculous rate, once again we owe this miracle to production ... following the crusades, industry developed enormously and brought to light a quantity of new mechanical (weaving, clockmaking, milling), chemical (dyeing, metallurgy, alcohol), and physical (spectacles) facts, and this not only gave enormous material for observation, but also itself provided quite other means for experimenting than previously existed, and allowed the construction of *new* instruments; it can be said that really systematic experimental science now became possible for the first time" (59a, 185). Thus the synthesis of empirical and theoretical activity, of abstract knowledge and concrete ability realised during the Renaissance signified the emergence of science in the proper sense.

Of course, it would be vulgar sociologising to interpret the maturing of scientific knowledge of nature as an immediate and direct consequence of the development of capitalism. In our view, this process (undoubtedly socio-cultural in nature) was determined by society in a more mediated and complex manner. Here is our idea of an adequate picture of the socio-cultural component in the genesis of the science of nature.

Natural science could only evolve as science under the conditions of capitalist commodity production, which gave an impetus to the axiological reorientation of cognition towards obtaining practically useful knowledge. However, this orientation was by itself insufficient for the formation of theoretical natural science, since, as we have already pointed out, orientation towards achieving applied results must be combined with the use of cognitive skills of working with idealised objects, with ideal modelling of reality.

To explain the social premises for preserving and developing these skills, it is not enough to refer to the capitalist mode of production.

The key to the causes for the preservation and further development of the activity, characteristic of antiquity, of constructing ideal objects, without which science is impossible, lies in the special significance of mediaeval culture, which played an exceptional role in this respect. Inasmuch as the formation of natural science necessitated a synthesis of abstract theoretical and experimental practical activity, and this synthesis, as we have established, could not have taken place under the slave-owning system of antiquity, it was necessary, at the initial stage, to retain the principles of working with idealisations while changing the system of production relations that stood in the way of this synthesis. Something like that had been realised in the early Middle Ages, of which the economic basis was no longer the slave-owning system but feudalism, while the intellectual basis was abstract theoretical activity involving ideal constructs (the theological speculative system of the world). The extremely specific conditions of mediaeval culture explain both the further advances in the "theoretical" study of nature and the absence of social bans on its "experimental" study (in alchemy, natural magic, etc.). In any case, the path from ideal modelling of reality to experiment was broken precisely in that period.

We can see just how difficult and far from simple that path was by considering that it took fourteen centuries for mankind to combine the abstract-theoretical (speculative-natural philosophical) tradition and that of the crafts.

Thus an essential extra-scientific premise for the formation of truly scientific natural knowledge was, along with the development of capitalist relations, the fact of assimilation within the framework of feudalism of the cultural traditions of antiquity. Taking this into account, the formation of natural science as true science, in terms of the socio-cultural determination of the synthesis of empirical and theoretical activity, can be reconstructed in the briefest outline as follows.

(1) The specific circumstances of the Middle Ages permitted the translation of the ratiocinative achievements of antiquity (the experiences of ideal modelling of reality) into the culture of the Renaissance, and the specific circumstances of the Renaissance permitted a substantive transformation of these achievements (this process, as we have pointed out, began already in the Middle Ages—there were fountainheads of experimental natural science in the monasteries)—an advance from the orientation towards the search for epistemological means of verifying the results of natural scientific research to

the formation of "technogenic" natural science. The transitional forms of the evolutionary chain leading from speculative natural philosophy to empirically substantiated natural science are such two-dimensional, empirical-theoretical phenomena as astrology, alchemy, natural magic, etc., as well as the theories of contemporary cultural figures such as Giordano Bruno, Roger Bacon, and others, which combined literally incompatible, in those times, empirical (experimental) and theoretical (theological-speculative) views and orientations.

(2) Later, consistent relegation to the intellectual periphery of fideistic, theological and metaphysical complexes (in the framework of deism) and the increasing tendency towards greater practical effectiveness of scientific activity (due to the progress of capitalist relations) gradually brought about a new and previously unknown intellectual phenomenon—theoretical natural science relying on experiment.

The assertion of the hypothetico-deductive methodology of cognition. The hypothetico-deductive method, the core of modern natural science—is based on logical deduction of statements from accepted hypotheses and on their subsequent empirical verification. The latter is seen as a procedure ensuring the establishment of the truth of theoretical assertions through their correlation with directly observed facts.

From this description of the hypothetico-deductive method underlying hypothetico-deductive theory one may proceed to a description of the latter. A hypothetico-deductive theory is a deductively formulated set of propositions, one that consists of syntax and interpretation. Unlike logico-mathematical (formal) systems, natural-scientific hypothetico-deductive theories are always interpreted, which means obligatory translatability ("projectibility") of their syntax onto a given fragment of reality (ontology) in relation to which the descriptive, explanatory and predictive functions of the theory are satisfied.

The hypothetico-deductive tactics of research was first introduced into science by Galileo. We refer here, first of all, to his theory of vacuum mechanics based on the principles of rational induction and mental experiment. To realise the essence of Galileo's innovations, we must expound, if only very briefly, Aristotle's science of nature, the critique of which stimulated Galileo's new programme for the construction of natural science.

Aristotle's physics includes a general theory of being which is, in modern terms, a concretisation of traditional ontology. Physical problems proper, in the modern acceptation, do not figure prominently in Aristotle's system, as analysis of the content of his few works on these problems shows. Aristotle's *Physics* comprises a general theory of nature, of the first elements and four causes. *De Caelo* deals with circular and straight-line, "natural" and "forced" motions. In the view of many historians, *Mechanical Problems* was written by Aristotle's epigones and not by Aristotle himself; these apocryphal writings mostly discuss technical problems whose solution relies on the uniform method of the lever.

The core of physical problematics in Aristotle is the theory of motion, which he originally associated with the concept of entelechy, or philosophical theory of actualisation. However, inasmuch as this interpretation of motion proved to be inadequate in the solution of particular physical problems, Aristotle was compelled to concretise it. With this aim in view he introduced more special concepts of types of motion (movement, change, growth, decrease), and later suggested an even more specific concept of change in the position of the body in the course of time (the concept of local motion), which he subsequently divided into natural and forced. To perceive the meaning of this distinction, we must characterise Aristotle's concept of space.

Space, according to Aristotle, is place, the boundary between the comprising and the comprised. The body within a comprising body is in a place. In accordance with the theory of the elements, the earth is in water, water in air, air in ether, and ether, in nothing. Thus, Aristotle's space, conditioned by the qualitative boundary between the object and the surrounding medium, is non-uniform and anisotropic. Hence the natural or forced character of motion is determined by the qualitative nature of its carrier. Thus fire moves up naturally, by its nature, and its motion down is forced, it is against nature; for the earth, the position above is contrary to nature, and so on. Since the motion of bodies is predetermined from the outset by the qualitative nature of their substratum, heavy bodies always move towards the centre, and light ones, towards the periphery.

Analysis of Aristotle's theory of non-uniform and anisotropic space affords a deeper insight into the essence of his mechanics. "In Aristotle's dynamics," writes Dierck-Ekkehard Liebscher, "no relativity exists between frames of reference, for the theorem of the conservation of impulses does not obtain. Forces are proportional not to changes of impulses but to the impulses themselves. The state of equilibrium of a force-free object is rest, which specifies a definite frame of reference.

In observing this state of equilibrium we can decide, for any reference system, what velocity it has in relation to the absolute state of rest" (162, 29).

What is the epistemological source of this position of Aristotle in physics? It is his crude uncritical empiricism and supernaive realism. In considering how bodies move in actual fact, Aristotle (a) failed to make an abstraction from the effects of friction, and (b) had to postulate the dependence of the velocities of motion on the qualitative properties of bodies and the characteristics of the medium.

It is this primitive physicalist approach that Galileo vigorously opposed, relying on the ideas of earlier critics of Aristotle. Already in his first work on the problem of motion (c. 1590), he criticised Aristotle's dynamics. In particular, Galileo refuted the Peripatetic doctrine of natural and forcible movements. He showed that where the medium in which motion takes place is not air but water, some heavy bodies (such as logs) become light, as they move upward. It followed that the movements of bodies up or down depended on their specific weight in relation to the medium and not on "predestination". In the same work Galileo showed the groundlessness of the Peripatetics' proposition that the velocities of the motion of bodies in less dense media are greater than in denser media—thus a thin balloon filled with air will move slower in air than in water, etc.

The positive part of Galileo's physical theory is expounded in his fundamental work *Discorsi e dimostrazioni matematiche*. Here, Galileo turned to the analysis of the isochronic character of the swinging of the pendulum. He concluded that pendulums differing in weight but identical in length perform oscillations of identical duration. But the movement of the pendulum is in fact the fall of a body along a curve of the circle. It follows that the force of gravity accelerates different falling bodies to an identical extent. Thus, if we neglect the resistance of the medium, all bodies in free fall must have the same velocity.

Simultaneously, Galileo conducted experiments in rolling bodies along an inclined plane, and here too he found confirmation of his idea of uniform acceleration of different bodies by the force of gravity. However, these experiments were not quite conclusive, since the action of the law of gravity was modified here by the action of external forces. To eliminate this defect, it was necessary to state clearly the nature of these modifications. The latter required a radical reformulation of the foundations of the prevailing Peripatetic dynamics adapted to the analysis of empirically recorded movements. Now, what course of action did Galileo choose?

He worked out a special type of research tactics which prescribed the study of ideal or theoretical motion described by the apparatus of mathematics, rather than of empirical motion. In accordance with this, Galileo's new dynamics fell into two parts, tentatively speaking. The first was intended to derive the laws of motion in pure form by logical deduction. The second, organically connected with the first, had to achieve an experimental verification of the abstract laws of motion obtained in the first part.

In developing his new dynamics, Galileo criticised the Peripatetic proposition "there is no action without a cause", which was only intended to cover the state of rest, in some such form: no body can move from the state of rest to the state of motion without application of some additional force. The Peripatetics believed that the cessation of motion was connected with the action of empirical conditions (friction, resistance of the medium) in case of cessation of the action of the motive force. Galileo introduced an essential correction in this interpretation: no body changes its velocity either in magnitude or direction without the action of some additional force. In other words, having once received an impulse, a body continues to move, when the action of the force has stopped, at a constant velocity regardless of the resistance of the medium and friction effects. This proposition revolutionised not only the field of science, signifying as it did the actual beginning of physics (the law of inertia), but also the domain of epistemology, as it destroyed Aristotle's naive physicalist views.

The point of departure of Galileo's physics is abstract and hypothetical. Aristotle described actually observed phenomena, while Galileo, logically possible ones. Aristotle considered the real space of events, and Galileo, relatively ideal space in which immediate research in the processes of nature was supplanted by analysis of mathematical limiting laws which could only be verified under exceptional circumstances. Characterising Galileo's epistemological method, students of his work point to mental experiment as a cognitive element which made an essential contribution to the arsenal of scientific activity. What is the essence of mental experiment, according to Galileo? The book of nature, Galileo believes, is written in the ideal language of mathematics. In reading it, one should resort to abstraction from the conditions of the empirical givenness of the processes under study, revealing the fundamental rational laws underlying sensuous appearance.

It appears natural in this connection that Galileo revived the epistemological traditions of Plato, who worked out an ideal-logical interpretation of the nature of knowledge. Aristotle had consciously broken away from Plato, rejecting his interpretation of the nature of knowledge, while Galileo substantiated the principle of intellectual rationalisation of the empirically given, the need to go to the essence beyond existence, and thus restored Platonism.

The view of the nature of cognitive activity in the spirit of Plato, as consisting of the study of limiting cases realised only under ideal conditions, is the new element introduced by Galileo, who thus added the method of mental experiment to the instruments of science.

The history of the evolution of the method of mental experiment, which so stimulated the formation of scientific natural theory, was in our view as follows.

(1) Results of real experiments (the side effects of the conditions of empirical realisability) naturally failed to confirm expectations—the share of negative results was too great. This caused attacks on Galileo's new theory of fall not only by his old opponents—the reactionary Peripatetics (cf. the criticism that came from the professors of Pisa)—but also by such progressive cultural figures of those times as, say, Descartes, who reproached Galileo for faulty experiments. Galileo found a way out of that dramatic situation in rationalising the experimentally obtained results. This enabled him to explain the negative results in terms of faults in the conditions of experimentation, or defects of the empirical level.

(2) Epistemological reflexion on the device, originally employed *ad hoc,* of rationalising the negative evidence in the experimental verification of theory, along with the conviction produced by that reflexion, that the interconnection between the theoretical and the empirical levels in scientific research is far from unequivocal, and also mediated in character, prompted Galileo the idea of a new method, the method of rational induction, which satisfied the conditions of artificial, abstract logical space, the space of ideal scientific reality, not of natural space. In this way the theory of vacuum mechanics crystallised: "If we were to eliminate completely the side effects of the empirical level, then—" (mental experiment).

(3) The development of the theory of vacuum mechanics logically culminated in the formation of the hypothetico-deduc-

tive methodology, since nothing but experiment could verify the ideal laws of motion deduced in vacuum mechanics. To be quite accurate, we must say that Galileo did not carry out his plan of empirical substantiation of the ideal laws of vacuum mechanics by comparing the ideal laws with the real ones (comprising a special system of amendments to account for empirical effect—friction, etc.). This plan was actually realised a century later, when the magnificent building of classical mechanics was completed.

In summing up the data on such highly important development as the assertion of the hypothetico-deductive methodology of cognition, we must stress the role of Galileo. It was Galileo who, by rejecting Aristotle's proposition that no motion can be continued ad infinitum (which was essentially equivalent to the discovery of the law of inertia, of which the precise formulation was, however, given only by Newton), laid the foundation of the science of nature. It was Galileo who undermined the naive qualitativist phenomenalism of the Peripatetics, revived the Platonist interpretation of the nature of knowledge, and worked out the research tactics of mental experiment in ideal reality, substantiating the possibility of employing in physics of the quantitative apparatus of mathematics, which signified its transposition onto a strict scientific basis. It was Galileo who stressed the need for consistent experimental verification of ideal logical laws and formulations and created a universal methodological framework for natural-scientific cognition.

That is why it is the figure of Galileo, the man who established the laws so clear and obvious now, created the very framework of thinking which made subsequent discoveries in science possible, reformed the intellect and provided it with a series of new concepts, and worked out a specific conception of nature and science (156)—it is this figure that marks the birth of truly scientific knowledge of nature.

The assertion of the principles of classical thinking introduced a whole series of new elements in intellectual life that have to be pointed out. Intellect was secularised and detheologised, and science was freed from the vise of the church, from the authority of canonical texts; analysis of the Holy Scripture gradually became the occupation of the monasteries, not the universities; academies became more and more the seat of science.

Scientific thinking was emancipated from fideist and organismic categories, rejecting the topographic hierarchy of top and bottom, which was central to the system of Catholic

Aristotelism. Spatio-temporal notions were desacralised as the ideas of uniformity and isotropy of space and time took shape and asserted themselves. Anthropocentrism was eliminated from science, and the picture of a unitary cosmos was accepted.

Scientific quest was made more democratic and efficient. Science gave up mediaeval dogmatism; the epistemological doctrines of those times, from Bacon's *Novum Organum* to Descartes' *Rules for the Direction of the Mind* and *Discourse on Method,* were highly critical and antischolastic in their orientation. St. Augustine's proposition "Believe, in order to understand" was rejected. The assertion of the progressist paradigm of scientific cognition discarded the scholastic authoritarianism of the *ratio scripta,* of a sacred, absolute and immutable "truth of the text", and did away with the mediaeval conception of the finality of the cognitive process.

Thinking came to rely on the foundation of causalism, on the paradigm of law-regulated and objectively existing nature permeated as a whole by natural causality and unitary laws. Science gave up the interpretation of concepts as independent elements acting as real universals, realising the need for experimental verification, of empirical control over discursively unfolded schemata and constructs.

The employment of measuring devices and operationalisation introduced the concepts of number and magnitude in knowledge and laid the beginning of exact science with its use of quantitative methods of analysis, calculation, processing, and evaluation of empirical data, which are amenable to mathematical modelling and subject to quantification. It was at that time that the hypothetico-deductive architectonics of natural scientific knowledge (the physics of principles) asserted itself, which made it possible to formulate quantitatively detailed and experimentally verifiable propositions.

The semantic structures that were necessary for the establishment of the mechanistic worldview as the dominant one became crystallised: supernatural individualising explanations in terms of concealed qualities responsible for the particular properties and behaviour of the phenomena under study were supplanted by natural explanations in terms of matter and motion permitting an interpretation of the essence of phenomena on the basis of the general principle of mechanic interaction of a substance with another substance; corpuscular notions, i.e., the view that all reality consisted of minute particles of matter, consolidated their positions; mathematically expressible and presentable concepts of "size" (extent) and "travel" (relative motion) took root as the basic sense-forming categories of thinking.

2.6. THE NATURE OF MODERN SCIENCE

Modern science is, epistemologically, a multidimensional phenomenon with numerous aspects. The cognitive complexes forming it are extremely polymorphous and belong to different levels. Modern science is a broad association of mathematical, natural-scientific, human and technical branches, of "disciplinary" and interdisciplinary studies, highly specialised and complex subdivisions functioning as discrete units of theoretical, empirical, formal, meaningful, fundamental, applied, and other kinds of knowledge.

At the same time there are grounds for assuming a certain "single axis", an essential unity of modern science, connected with the specificity of research strategy, the style of the formulation and study of problems, the mode of the production and functioning of knowledge, the nature of prospecting activity, etc.—in short, with everything that constitutes the specificity of the total potential of science fixed in terms of epistemological analysis.

Methodological works that have appeared in the last twenty-five years often stress the inner affinity of different types of knowledge integrated in modern science, even those that belong to the rigorous natural sciences and non-rigorous social and human sciences.

To find out the nature of that community, and to identify the grounds that unite factually diverse phenomena in the single whole that we call modern science, it is not enough to carry out a functional analysis of its synchronously active structures. We must also employ the instrument of comparative analysis permitting a typological juxtaposition of modern science and the epistemological structure genetically preceding it.

This comparison shows the following.

In speaking of the need to correlate modern science with the structure of the same order which historically precedes it, and in which its direct sources lie, we refer to classical science. Its critique and reinterpretation of the cognitive norms and ideals specifies, properly speaking, those inner basic relations which fully determine the epistemological features and constitute the conditions of integrality of modern science. Classical science is here taken to mean quite a definite research and cognitive culture which was realised as a predominant tendency between the 17th and early 20th century, at the end of which period the quantum-relativist epoch began.

In the framework of a typological comparison which interprets the transition from classical to modern science as crystallisation of a different research culture corresponding to a new spiritual formation rather than as a mere shift in problems and subject matter, in the experimental and technical equipment, fundamental stylistic features of classical science must be pointed out which distinguish it from modern science, which make it a separate cognitive epoch and a stage in the development of the scientific intellect.

Classical science functioned as an entirely integral structure; the foundations of this integrality were determined by a series of substantive orientations. Of these, let us especially single out two.

(1) Orientation towards a final-objective system of knowledge embodying truth in its final and accomplished form. Based on classical mechanics, which was regarded as a universal method of cognition of the world's phenomena and at the same time as a standard for any science, this orientation was supported by a whole series of particular tendencies.

(a) The tendency towards single-valued interpretation of events; exclusion from cognitive results of chance and probability seen as indication of incompleteness of knowledge or subjectivism.

(b) The tendency towards eliminating from the context of science the characteristics of the researcher, which were alleged to interfere with adequate identification of the truth; rejection of the need for taking into account the specific features (modes, means, and conditions) of the subject's cognitive assimilation of the object.

(c) The attempts at establishing the substantionality and the primitive elements of the world.

(d) The tendency to regard knowledge forming the actual body of science as absolutely reliable and non-problematisable. This feature was, of course, fixed in philosophical-methodological consciousness, which founded science on the proposition that "there is only one truth about each thing, and whoever finds it, knows about it as much as anyone can know" (132, 15).

(e) The tendency to interpret the nature of cognitive activity in terms of the naive realistic correspondence concept postulating a mirror-like and immediately obvious harmony between knowledge and reality, that is to say, uncritically accepting the dogma that everything cognised as belonging to a thing is in actual fact an attribute of that thing.

(2) The view of nature as a whole immediately given from the very beginning, always equal to itself, and devoid of development, going round and round along the same eternal and limited circles (59b, 5, 39). This orientation was concretised in such research stratagems, specific for classical science, as emphasis on stationary states, elementarism, and anti-evolutionism.

The efforts of classical scientists were mostly aimed at the identification and definition of the simple elements of complex structures, while the complex functional-genetic links and relations existing within these structures as dynamic wholes were obviously and consciously ignored. The interpretation of the phenomena of reality was therefore entirely metaphysical, i.e., devoid of the perception of their contradictoriness, mutability, transformability, historicity, etc. Suffice it to mention in this connection the following principles, typical of classical science and fully reflecting and expressing its ideological aspirations as the principles of constancy[4]: the principle of mass constancy (Newton), the criterion of the constancy of the composition of a chemical compound (Proust), the proposition concerning the qualitative and quantitative immutability of the organic species after their divine creation (Linnaeus), etc.

What has changed since the classical epoch, what marks the entry of science in the non-classical phase of its development? A great deal has changed, but we are only interested in the epistemological aspect of all these changes.

The transition from classical to non-classical (modern) science, and the changes it produced in the objective content of knowledge, in its foundations (the modes of the analysis of objects, in obtaining, developing, and structuring of the ingredients of science), in the type of the self-consciousness of science itself, have been called a "revolution". In brief, the essence of this revolution may be described as follows: it was produced by one single factor—the entry in the body of knowledge of the subject of cognition, of his activity, as a necessary and inalienable component. It would be hard to exaggerate the fundamental importance of this circumstance.

The paradigm of classical science, with its orientation towards the cognitive assimilation of the object in itself, so to speak, in its essential naturalistic immediate givenness, raised to an absolute the concept of natural process specified regardless of the conditions of its study. This entailed the familiar elimination from science of subjective activity and neglect for the role of the researcher's instruments making an impact on the object of cognition. Unburdened by reflexion on the subject's specific functions in a cognitive situation, the ideologues of classical science cultivated the dogma that it was possible to specify any cognitive parameters without any limitations and to detail them in all aspects.

The revolution in science which destroyed the illusions concerning absolute knowability of processes studied and the possibility of their cognitive detailing in all aspects signified, as has been pointed out above, the replacement of the contemplative style of thinking by the activity-oriented one. Let us consider some of the implications of this proposition.

(a) The incorporation of subjective activity in the context of science led to a change in the perception of the object of knowledge. The latter is now perceived not as reality "in pure form", as given in living contemplation, but as a kind of cross-section specified in terms of accepted theoretical and operational instruments and modes of its assimilation by the subject. It is meaningless to speak of the characteristics of objects without referring to the instruments used to determine these characteristics, and modern science therefore has accepted the relativity of the properties of objects, which depend on the type of their interaction with these instruments in cognitive situations.

(b) The awareness of the dependence of the object on research and transformation stimulated the transition of science from "the study of things regarded as immutable and capable of forming certain connections to the study of conditions under which a thing does not just behave in a certain manner but can be or not be something, can exist or not exist as a given definiteness only under these conditions" (88, 73). For this reason, modern scientific theory begins with the specification of the procedural basis, with identifying the modes and conditions of the study of the object, which forms the semantic and operational outline of the theory, safeguarding the objectiveness and harmoniousness of the description of the facts it describes.

(c) The dependence of the picture of the object on the relation to the instruments of cognition, and the consequent need to organise knowledge with proper regard for the real operational procedures, determine the special role of the measuring devices (or experimental apparatus) in modern scientific cognition. Without such devices, it is sometimes impossible to identify the object of science (or theory), as it is only identified in the interaction between the object and the device.

(d) The interaction between object and device, which justifies only the analysis of concrete manifestations of the aspects and properties of the object at different times and in differently realised situations, cannot but result in a kind of spread, entirely objective, in the final results of research. This forms the basis of complementarity, in the broad sense; indicating the various manifestations of the properties of an object depending on the type of its interaction with the device under different, often mutually exclusive conditions, such complementarity shows that different types of the description of the object, its different conceptual images, are all equally justifiable. Properly speaking, this explains the fact that modern research activity has moved from the single infinite "object in general with an unambiguous and immutable" "nucleus", an object "reflected in the only possible true manner", to a "world reminding one, rather, of a kaleidoscope of a great many projections" (ibid.), to a world described in a system of finite pictures related to, and dependent on, the instruments of cognitive assimilation, none of which can claim to represent an allround and comprehensive type of description.

(e) The rejection of the contemplative spirit and naive realism of classical science reflected, in particular, in the new practice of specifying the object of knowledge with due consideration for the mode of its cognitive assimilation, in the understanding of the dynamics of the links between the empirical and the theoretical, etc., has changed the status of fact as verification instance. What we mean here is this. The increased dynamic element in science—greater mathematisation, the merging of fundamental and applied research, expansion of the quest from the domain of the real to that of the possible, the study of extremely abstract types of reality absolutely unknown to classical science—potential ones in quantum mechanics and virtual ones in high-energy physics—and so on, has resulted in a sort of mutual penetrability of fact and theory. This mutual penetrability sometimes, as, e.g., in the case of resonances, assumes such unexpected and quaint forms that the boundary between the empirical and the theoretical is difficult to draw, and the familiar line between fact and theory disappears. In this connection, the conception of verification experiment has changed. Firstly, it is no longer able to act as a separate judge of theory, and is now realised, as an epistemological procedure, as part of a package together with other modes of verification of knowledge—the intratheoretical ones, such as the principle of correspondence, the establishment of the inner and coherent perfection of theory. Secondly, it no longer shows, primarily, that a given theoretical formulation corresponds to something that exists before the experiment "in actual fact", but rather the fact that a theoretical assumption is justified for given conditions.

Such is, in the briefest outline, the nature of the non-classical style of thinking asserted by the scientific revolution. A detailed exposition of its epistemological consequences necessarily covers the question of the essential transformations of the traditional classical norms and ideals of research. We shall restrict our analysis of this question to such fundamental epistemological concepts as the modes of introducing objects in theory, the ways of solving the problem of existence, the character of activity at the empirical level, and the criteria of exactness and rigorousness that set the objectives of scientific research.

At the classical stage of the functioning of science, when meaningful and conceptual clarity preceded the full comprehension of mathematical structures, the universal mode of specifying the objects of a theory were the operations of abstraction and immediate generalisation of the available empirical material (the theory of abstractions of classical philosophical empiricism adopted by the scientists of the classical period, which in its turn had crystallised as a generalisation of their research tactics). The problem of existence, central to theoretical activity, was also solved in the same naively realistic key. Immediate obviousness was seen as the decisive condition of existence: only that which is given in intuitively intelligible perception exists.

The naively realistic conception of cognition accepted in classical science can no longer serve as a guiding principle for modern science, in which the instrument of specifying theoretical objects is mathematics, the share of auxiliary models mediating the connections between theory and reality increases, conceptual clarity no longer precedes the perception of mathematical structures, and science finds it even more difficult to grasp their content.

In modern science, objects are introduced on the paths of mathematisation (at any rate, there is a tendency towards that) interpreted as extensive introduction of the arsenal of mathematical instruments in concrete scientific cognition, necessitated both by the needs for effective elaboration of the conceptual basis of sciences (mathematisation as a means of theoretisation) and by the desire to optimise the forms of the organisation of knowledge (mathematisation as a means of formalisation, of overcoming the qualitative and non-rigorous character of knowledge).

In classical science, which exploited the resources of mathematics fairly intensively, mathematisation was obstructed by the pervasive and very strong desire for direct perceptibility, for empirically or intuitively verifiable obviousness. That is why a qualitatively new stage in the mathematisation of science began with the formation of non-classical science which decisively rejected that tendency blocking the process of mathematisation; this immediately opened up the possibility of operating with highly abstract structures which, as a rule, had no prototypes in intuition or direct observation. From this moment began the universal application in natural science of the methods of mathematical hypothesis, the hypothetico-deductive method, the method of the principles of theory construction, and so on, which marked the universal tendency of mathematisation—a very powerful generator of ideas in science leading to the emergence of new scientific branches and theories, of which the construction of the general theory of relativity is an excellent illustration.

The most fundamental epistemological consequence of mathematisation, as we have pointed out, is a higher level of the abstractness of science, the loss of intuitive clarity in the perception of the object. Following Descartes, classical scientists traditionally regarded mathematics as an embodiment of intuitive obviousness, but in actual fact it contained propositions that were far from obvious. Suffice it to point out in this connection the well-known vagueness of the formulation of Euclid's fifth postulate (attempts to prove it culminated in the construction of non-Euclidian geometries); or the problems of the foundations of Cantor's set theory, the reasons for the defectiveness of which are still not clear. We can also cite the lack of clarity in the conceptual status of the continuum hypothesis, the axiom of choice, and so on. Thus modern science can no longer rely, in principle, on the epistemological precepts for introducing theoretical objects which were accepted in classical science.

The same applies to the strategy of solving the problem of existence. Having given up the classical criterion of intuitive obviousness, modern science proceeds from the mutual complementarity of the weak and strong interpretations of the foundations of existence. The former accepts that it is possible to admit the existence of something on the grounds that there is nothing in available knowledge to forbid it. This interpretation is restricted by the available stock of knowledge (which can be extended as further laws prohibiting the existence of something are discovered) and asserts potential existence only.

The second version is guided by the criterion of experimental verifiability, discoverability, etc.; it asserts actual existence. It is easy to see that the weak version embodies the necessary conditions of existence, while the strong version, the sufficient ones; therefore many theoretical entities like quarks, planckeons, etc., satisfying the weak but not the strong condition of existence are not regarded as actually existing. It is appropriate in this connection to stress the role of the principle of experimental verifiability which is seen in modern science as a fundamental one; it is not intuitive obviousness but experimental adaptedness that makes modern scientific concepts meaningful.

In modern science, the ideas of the principles of activity at the empirical level of research have also changed in consequence of the replacement of the monofactor experiment by the multifactor one.

The basic concept in the theory of monofactor experiment in classical science was that of stabilised morphology of the object, whose nature of functioning was considered in isolation both from the constituents and their surrounding complexes. Practical development of experimental activity increasingly revealed, however, the uncritical character of this classical theory, which assumed that it was possible to analyse separate parts of objects or processes existing independently. It became necessary to discard the inadequate classical conception of the monofactor experiment, and it was replaced by the non-classical conception of multifactor experiment which accords with the new practice of experimentation in science.

The essential feature here is that underlying this theory is the interpretation of objects as complex dynamic structures constituting self-changing systems.

This conception is thus based on a different view of the object's ontology comprising the following extremely important orientations that introduce new elements in the perception of the object's possible cognition.

(a) Rejection of the isolation of the object from surrounding influences (allegedly intended to ensure the purity of analysis).

(b) Recognition of the dependence of the definiteness of the object's properties on the dynamic and complex nature of its functioning in the cognitive situation.

(c) A systems-integral evaluation of the object's behaviour, the latter being seen as conditioned both by the logic of internal change and the forms of interaction with other objects.

(d) The dynamisation of the ideas of the essence of the object—the transition from the study of equilibrium structural organisations to analysis of non-equilibrium, non-stationary structures behaving as open systems. This conditions the researcher to study the object as the focus of complex feedback links resulting from the action of various agents and counter-agents. Thus in chemistry the assertion of the dynamic approach revealed the special role of catalysts; catalysis, once a method of chemistry, became the subject matter of analysis (catalyst chemistry); as a result, such non-classical concepts emerged here as chemical time, chemical evolution, etc.

(e) Anti-elementarism. Interdisciplinary studies of dynamically functioning open non-equilibrium systems resulted in the rejection of the orientation towards identifying "elementary constituents" (especially in biological, sociological, economic, space, agrobiological and other systems, and in design, prognostic, and other types of tasks). In modern science the concept of elementarity is relativised and no longer regarded as absolute, as in classical science.

The concepts of the criteria of exactness and rigorousness that are the guidelines of the cognitive quest have also changed in modern science. These criteria are now taken to mean a set of rules for incorporating in the body of science of logically substantiated and quantitatively detailed propositions. Let us stress that these rules are inseparable from science as a rational and demonstrative type of theoretical assimilation of reality and are immanently inherent in it. In this case, it is a question of a better logical or experimental substantiation of scientific knowledge, of increasing the standard of exactness and rigorousness.

In the classical period of the functioning of science, however, the desire for exactness and rigorousness, always inherent in the minds of scientists, was uncritically hyperbolised. Only knowledge that was substantiated in all aspects was believed to be scientific (the Laplacean ideal in methodology). Accordingly, the presence of probability was seen as a sign of insufficient substantiation, of problematic character of the units of knowledge; knowledge that was not absolutely clear was not believed to be true knowledge and was therefore automatically excluded from science.

In the course of time, however, the situation changed significantly—in the following aspects.

The proofs of the propositions of modern science, which are obtained by collectives rather than individuals and require the wide use of computers, undoubtedly become less obvious, accessible to direct perception, etc., and in this sense less exact and rigorous from the standpoint of classical science. But that is not the main point. The main point is that, as the historical experiences of the development of science have shown, the notion of "absolute" exactness and rigorousness in the evaluation of cognitive results is, generally speaking, meaningless. To prove this, let us pose such a question: to what limit can the exactness and rigorousness of knowledge be increased? It appears that this process is not unlimited. Where the instrument of achieving exactness and rigorousness is formalisation, the general methodological limitations of Gödel and Tarski obtain. Where the instrument is experimentation, certain specific general and particular limitations apply. The particular limitations include the resolution of the apparatus employed, in view of which a certain fixed level of exactness and rigorousness of research cannot be exceeded. General limitations comprise the quantum restrictions as stipulated by the content of the principles of complementarity and indeterminacy.

The necessity of taking into account these limitations determines such a feature of modern science as constructivity, which is realised as a paradigm of the non-classical conception of the ideals of exactness and rigorousness.

Another element that is worth mentioning here is the cyberneticisation and automation of the calculation basis of science. The point is that in calculations done by computers, which have become an inalienable element of modern scientific activity, "the basic numerical data are always specified to a certain finite magnitude", which introduces an "ineliminable error of solution, which can be significant, whatever the accuracy of the calculation" (17, 92). All this also destroys the classical ideal of exact and rigorous knowledge of which the detailing was supposed to be quantitatively unlimited; all this conditions the non-classical "inexactness" and "non-rigorousness" of modern science.

Yet another element that eroded the classical ideal of exact and rigorous knowledge is connected with the identification and study of the so-called incorrect problems, i.e., problems for which even the concept of approximate solution is actually meaningless, as these solutions can widely vary with the slightest changes in the initial data. The psychological effect of the discovery of incorrect tasks was so great that Jacques Hadamard, who contributed a great deal to their study, refused to recognise them as justifiable and acceptable (the so-called Hadamard ban).

After Andrei Tikhonov worked out a general method, the method of regularisation, for the solution of incorrect problems which arose in geophysics, geology, seismology, medicine, economy, astrophysics, radiolocation, etc., the impression was that the concept of "approximate solution" was restored to its original status in science. Further study has shown that the explication of the problem of regularisation, which is central to the solution of the whole issue of incorrect problems, rests on the explication of the general problem of Banach's analytical representability, which is as yet unsolved and the solution of which is linked with future achievements in the study of set theory axioms.

It is clear at the same time that the concept of approximate solution is relative, for it became clear that "a problem may be regularisable in some types of mathematical spaces and unregularisable in others" (17, 93).

Thus, modern science is guided by the idea of relativity of the criteria of exactness and rigorousness, which increases the distance between modern science and the interpretation of the nature of these criteria in classical science.

Modern knowledge is permeated by the spirit of historicity; the assertion of this spirit is one of the most significant results of the scientific revolution. In brief, the new elements introduced by historicity, which deepened the difference between non-classical and classical science, are as follows.

(1) There has occurred a change in the relation of science to the knowledge that makes up its body. The scientific revolution has demonstrated the relative status of both knowledge and of the foundations of knowledge; it put an end to the classical myth of the immutability of the principles of science, and freed its self-consciousness from the dogma of the continuously progressive nature of scientific development as cumulative detailing of a few fundamental "a priori self-obvious" ideas. All this, raised to the rank of the paradigm of modern science, determined its healthy self-criticism and the idea of the need for continual revision of its principles and assumptions as well as algorithms for obtaining results.

(2) The logical foundations of science and knowledge in general have also greatly changed. The essence of the changes lies in the fact that science uses a logical apparatus which

is best suited to reflect the specificity of the activity-oriented approach to the analysis of the phenomena of reality. That is the underlying factor in the progress of non-classical multi-valued logics; the limitations on and rejection of such classical logical devices and norms as the induction theorem, the rule of excluded middle, etc.; and the introduction of the concept of the finite and constructive nature of both the objects of science and the operations on these objects.

In conclusion let us define the tendencies which, in our view, will largely determine the directions of the expected progress of science. We are not concerned here, of course, with compiling projects for the future development of science but with stating the actual vectors in the dynamics of knowledge which are gathering momentum in the present and will be fully manifested in the future, being reflected in one way or another in the structure of future science.

Let us begin with the tendency towards interdisciplinary study which, in our view, will be the leading tendency in the formation of the domain of future science. As used here, the term "interdisciplinary" refers to the obliteration of boundary lines and partitions between the traditionally professionally isolated natural, social and technical sciences, as well as to the increased centripetal processes within science resulting in the intensification of the relations of unity, interconnectedness, and mutual penetration and interaction between different scientific subdivisions termed scientific disciplines.

The tendency towards interdisciplinary research, which gave rise to the development in modern science of general systems theory, ergonomics, systems engineering, design, the theory of automata, ecology, etc., is determined by the impossibility of positive solution of the problems which science faces (owing to the complexity and multiaspectual character of these problems) in the framework of the separate systems of knowledge without resorting to the conceptual, operational, etc., resources of other cognitive fields, without creating a cooperative research organisation. The practice of interdisciplinary cooperation of research will naturally produce new forms, norms, standards and ideals of activity in science; it is easy to predict that a characteristic feature of this future scientific activity will be the impact of these norms and ideals on the scientist, who will have to take a systems position within interdisciplinary research rather than lose himself in a single narrow professional role which would take such a scientist beyond the interdisciplinary research. In this connection the role will certainly increase of generalising reflexion as an instrument of specifying the unity of the subject-matter and methods of the new, systems type of scientific activity.

Changes in the sphere of experiment will be largely determined by the development of the modes of overcoming of those actual difficulties which will arise in experiments involving complex dynamically functioning systems of integral objects. A fundamental difficulty here is the absence of invariant conditions in the reproduction of the experiment, in copying it. The interdisciplinary and systems nature of experimenting—the involvement in the experimental situation of a large number of factors, and the change in the situation itself due to the statistical character of the interaction between agent and counteragent—brings the scientist face to face not only with objective spread of the initial and attendant conditions, which are mostly, because of their multifactor character, very hard to control but also with considerable spread of the final results of the experiments in a given series. The latter circumstance deforms the criterion of reproducibility, according to which identical results must be obtained under identical circumstances.

To overcome this difficulty, it is necessary to create, first of all, a new mathematical apparatus suited to the description of nonstationary processes, which will naturally stimulate advances in the theory of nonlinear equations, in statistical methods of research, in the methods of calculation and prognostication of dispersion, in the methods of taking into account the kinetics of processes inseparable from the processes themselves, etc.

Along with this, the role will increase of mathematical and simulation experiments in situations in which it is important to exclude possible morphological changes in the object; such experimentation is a constructive, idealised, conceptual study or modelling, design, etc., of the behaviour of an object long before the potential real experimentation. The fact that mathematical and simulation experiments are not connected with any concrete conditions of experimenting, which accounts for their great mobility and effectiveness, will result in their wide employment in science and in consequent narrowing down or complete disappearance of the sphere of application of full-scale experiments. This will especially apply to such fields as ecology, economy, genetics, demography, geology, astrophysics, geography, high-energy physics, catalyst chemistry, space research, etc.

The future of full-scale experiment is, we believe, connected with the wide introduction in science of the methods of long-base interferometry, projection television, high-speed holography, photonics, videography, direct-shadow methods, and numerous latest optical observation methods adapted to the recording of fast processes. All this will open up the possibility of "conducting a posteriori analysis of a recorded process during the time required without loss of amplitude and phase spatial and temporal information".

The role will increase of classification and systematisation as means of ordering scientific information. The content of the principles of symmetry, relativity, invariance, as well as the laws of exclusion and conservation, will expand. If it proves impossible to invoke these laws (where the rule applies, e.g., that the weaker the interaction the more principles of symmetry it violates), new principles of objective fixing of reality in theoretical form will be worked out.

The significance will grow of planning and prognostication, and of evaluating the effectiveness of knowledge; this activity will specialise in assessing the prospects for the progress of scientific disciplines and in finding the shortest ways of introducing theoretical achievements in industry and technology.

Allround computerisation and mechanisation of knowledge will give rise to an even greater relativisation of the criteria of exactness and rigorousness. Computer programmes specifying only the general algorithm of solving the tasks of a given class do not contain an analysis of the solution of concrete problems, as these are too numerous. Hence the inevitability of errors in the application of general programmes to particular cases.

The share of socio-political, moral-ethical, humanist and similar reflexive evaluations, realised as modes of inner self-control of knowledge, will increase.

The extremely rapid progress of science will probably reveal in the nearest future many important aspects in the conception of the nature of science itself. The spirit of changes will therefore pervade the methodological sphere as well.

Chapter 3

FORMS OF SCIENTIFIC KNOWLEDGE

Our inquiry into the formation of science has elucidated only some of the problems we intend to discuss. Other questions, quite a few in number, are connected with the study of an already evolved science as a system of knowledge—heterogeneous yet retaining epistemological unity. These questions will be dealt with in the present chapter. Analysis will be focused on specific features of the familiar forms of scientific cognition: mathematics, natural science, the human sciences, and technological sciences.

3.1. MATHEMATICS

The science of mathematics is an association of deductive theories (arithmetics, algebra, geometry, etc.) reflecting certain fixed domains (those of numbers, functions, spaces, etc.). They can be divided into pure and applied. This division is of course conventional and relative.

First, various areas of mathematics are marked by a theoretical-methodological unity and interpenetration. For instance, applied mathematics had as its source fundamental research; certain abstract concepts—the axioms of set theory—are used in applied mathematics in the construction of numerical algorithms.

Second, the concepts of pure and applied are mobile. Mathematical logic, which was only recently regarded as part of pure mathematics, is now seen as a constituent element of applied mathematics; without it, computing mathematics would be impossible.

Pure mathematics comprises abstract theories functioning as the conceptual apparatus of mathematics proper (e.g., analysis and algebra) and as a means of substantiating mathematical theories (set theory, metamathematics).

Applied mathematics forms the foundation of computer and microprocessual technology and of robotics, of the entire ensemble of electronic technology; it also links up mathematics and other sciences, practice, and production (programming, control theory, reliability theory, etc.)

The concept of mathematical theory as a form of thinking can be made more specific by introducing the concept of relational system S comprising set μ of arbitrary objects connected with each other by definite relations R_1, R_2, ... R_n (62, 91), or, in more compact symbolic notation,

$$S = \langle \mu, R_1, R_2 \dots R_n \rangle$$

Apart from mathematical theories proper that constitute the body of mathematics, it also includes the apparatus of logic L—the framework or skeleton of mathematics that lends it the status of a deductive science.

In view of this, the structure of mathematics can be adequately expressed by the formula $\langle \Sigma S; L \rangle$.

Mathematics has the following specific epistemological features.

The absence of direct (rigid) correlation with any fixed fragment of reality, which makes mathematics more abstract than other branches of knowledge.

In set-theoretical terms, so popular these days, mathematics studies formal relations between certain classes of sets regardless of their factual, "material" nature.

Analysing ontologically unspecified systems, mathematics studies abstract structures.

We find a detailed development of this approach in Nicolas Bourbaki. They explain that, to define a structure, certain relations between the elements of a set are specified, and then the postulate is accepted that these relations satisfy certain conditions; the conditions are listed and defined as the axioms of the structure in question (121). Then logical corollaries are deduced from the axioms of the structure which form a mathematical theory unconnected directly with reality (as there is no place here for notions about the nature of the elements described).

Genetically, a modern mathematical theory is constructed as a unification of the axioms of a basic theory (a system of logic or arithmetic) with a certain special axiomatic system. Thus the axioms of a number of structures of Euclidian geometry are obtained by adding "the Kolmogorov axiom to the axioms of the predicate calculus containing the relation of equality and the axioms of set theory" (31, 62).

To identify the referents of the abstract properties and relations studied in matematical theories, these theories are given an interpretation. A set of interpretations or models forms the meaning of a mathematical theory. If the class of models is empty, the theory is meaningless or contradictory.

Axiomatism. The fact that Euclidian geometry, built on the basis of the genetic-constructive method, embodied, during a very long period of time, the essence of mathematics, made a considerable impact on the methodological aspects of the cognitive process in this science. In fact, the view prevailed that the Euclidian method was the only possible method of constructing a mathematical theory. Thus, Leibnitz, who extended the ideal of geometry to embrace the whole of mathematics, believed that its specificity did not lie in logical proof but in immediate perceptibility, which he believed to be theoretical and not sensuous in nature.

Leibnitz's idea of inner contemplation, which registered, according to the prevailing view, the specificity of cognitive activity in mathematics, was adopted and further developed by Immanuel Kant. In his handling of the question of the possibility of theoretical mathematics, which he associated with the problem of the possibility of synthetic a priori propositions, Kant started out from the typology of demonstrative vs. discursive proofs. He distinguished between mathematical proof, which in his view appealed to immediate perception, and which he called demonstrative, and philosophical proof, conducted on a conceptual-verbal basis, which he termed discursive. It is easy to see that the prototype of demonstrative proof, which Kant raised to the rank of a universal, and which he substantiated by means of an extremely cumbersome transcendental aesthetics purporting to explain the intuitive obviousness of mathematics, was the entirely real genetic-constructive method of expounding geometrical knowledge realised by Euclid. Using a historically concrete actual form of scientific knowledge, Kant raised it to an absolute, making it a suprahistorical a priori form and a condition of the existence of mathematics in general.

Thus, the view that the specificity of mathematical cognition lay in combining "demonstrative" structures of inner contemplation entailed the positing of direct perceptibility as a universal methodological regulator. Not only proofs but the very propositions of mathematics, axioms first and foremost, were evaluated in terms of this regulator. This explains, among other things, the familiar and extremely inadequate interpretation of axioms as limiting heuristically powerful propositions distinguished for their self-obviousness. It took a great deal of time and factual evidence of restructuring and clarification of systems of mathematical knowledge to bring the conviction that the truth of axioms or the basic postulates of a mathematical theory is far from self-obvious. The restructuring of systems of mathematical knowledge that is still in progress (thus algebra has gone through three transfigurations during the present century) shows that the clarity, truth and reliability of axioms must be established in a practical manner, through moving from the domain of theoretical activity into that of creative transforming practical activity, rather than through appeals to self-obviousness.

The proposition that mathematics is mediated by practice was recently questioned by the proponents of the logical programme in the foundations of mathematics. However, careful analysis of that programme showed that many mathematical axioms could not be given a purely logical (analytical) substantiation—recourse must be had to the extralogical sphere. Thus, Bertrand Russell had to introduce the axiom of infinity as he evolved his theory of types. At first he believed that that axiom was not a secondary assumption different from the assumptions proved on logical grounds alone. It was later demonstrated that the axiom of infinity "is not analytic according to any known theory of deductive reasoning" (127, 183). The axiom of reducibility offers a similar case. The principles of constructing a mathematical theory demand that the specificity of the basic theory should be taken into account, and this is linked with evaluating the specificity of meaningful assumptions (ontologies) forming its framework and belonging to the domain of the factual; in this way the (mediated) joining of mathematical theory, practice and reality is achieved.

We can thus draw the conclusion that the universalisation of the criterion of direct perceptibility and, of course, the concomitant neglect for the criterion of practice in substantiating mathematics, signified a psychologist and subjectivist approach to mathematics. It was this trend in the methodology of mathematics that was rejected by Bernhard Bolzano who especially criticised Kant's identification of the demonstrative method of cognition with the mathematical one.

The demonstrative mode of cognition, which appeals to construction, perceptibility, contemplation, etc., is not, in Bolzano's view, a method of real proof at all—it is merely a means of explicating provable propositions. Truly rigorous proof and substantiation are embodied, according to the author of *Wissenschaftslehre*, in pure discursive proof in which demonstration is avoided thanks to the operation of logical ascension to the objective foundations of the truth.[1]

The significance of Bolzano's critique of the genetic-constructive method and elaboration of an apparatus of logical proof lay in the fact that it showed the need to construct mathematics on the principles of a logically rigorous deductive science.

Another line in the critique of the ideal of Euclidian mathematics developed in the direction of a re-interpretation of the notions of the subject-matter of mathematics. Inasmuch as the axioms of Euclid's geometry were meaningful, and geometry itself was seen as describing and relating to empirically given space, the universalisation of the epistemological status of those axioms entailed the interpretation of mathematics (geometry) as physics. Later this position was somewhat modified, but the principle itself of rigid links between the subject-matter of mathematics and one of its possible interpretations remained intact; mathematics was interpreted as the science of number or quantity in general, if not as physics. Attempts to free the sphere of mathematical cognition from its derivative interpretations have been made since the times of George Boole, who made this statement of methodological significance: "It is not of the essence of mathematics to be conversant with the ideas of number and quantity" (118, 13).

Boole's ideas were shared by Hermann Grassmann, who constructed an abstract mathematical system free from rigid association either with the concept of number or that of geometrical object. Julius Dedekind was also inclined towards an abstract logical interpretation of the subject-matter of mathematics. The position linking up the subject-matter of mathematics with one of its possible interpretations finally collapsed in the late 19th century, when research in the foundations of mathematics led to its arithmeticisation. Augustin Cauchy, Karl Weierstrass, Dedekind and Georg Cantor showed that mathematics (algebra, analysis, the theory of functions), along with analytical geometry, can be built on the basis of arithmetic of natural numbers, which came to be regarded as the foundation of mathematics. The task of axiomatisation of arithmetic was soon carried out by Dedekind and Peano. In this way the efforts of these scientists, who reduced mathematics to arithmetic interpreted as an abstract axiomatic theory, provided the basis for the requirement of the purity of mathematics in the sense of the absence of unambiguous association of its subject-matter (content) with an interpretation specified a priori.

To sum up. When the two lines of the critique of the Euclidian ideal of mathematics (one deriving from Bolzano, the other from Boole) merged, an entirely new view of the nature of mathematical activity, based on the axiomatic ideal, emerged. The programme for the construction of mathematical theory on a rigorous axiomatic basis, which is the general norm in modern mathematics, was worked out, in the field of geometry, by David Hilbert, who refused to ascribe any concrete physical images to the fundamental concepts of geometry. Hilbert introduced axioms that had no interpretation whatever. He insisted, for instance, that if the geometrical terms "point", "straight line", "plane" were to be replaced by such terms of everyday language as "table", "chair", "beer mug", geometrical theory in itself would be neither better nor worse (144); the criterion of its truth would be logical consistency and deducibility from the axioms. Thus we see that Hilbert firmly insisted on distinguishing clearly between theory and interpretation of theory.

A formalised theory, according to Hilbert, permitted a number of interpretations, as its propositions were not directly correlated with reality. Before an interpretation is provided (a spatial one as in Euclid, or an arithmetical one as in Felix Klein, in the case of geometry), an abstract deductive system is therefore no more a geometry than any other theory.

The desire to axiomatise and to formalise the system of knowledge stems from the fact that (a) it is impossible to use effectively the apparatus of logic in a non-formalised system; (b) it is not always clear if such a system is complete; if it is not complete, it is not clear whether it contains premises which can give rise to contradictions, under definite circumstances.

Exactness and rigorousness. The following are the underlying causes of exactness and rigorousness.

(a) The apodictic nature of proof, resulting from the axiomatic-deductive organisation of mathematical knowledge.

(b) The algorithmic nature of proof, interpreted as the existence of fixed modes of solving mathematical problems in the form of systematically deduced unambiguous instructions.

(c) The deductive nature of mathematics, embodied in the principles of constructing discourse in it, based on the transition from one meaningful (informational) structure to another according to clear-cut and rigid rules of logic. The deductiveness of mathematics signifies the existence of a logical path from axioms to theory, and that path is logical inference—a concentrate, as it were, of the essence of mathematics. In view of this, parallels can be drawn between science and mathematics (science is discourse, discourse is mathematics, *ergo*, science is mathematics).

(d) Unambiguous definitions. In accordance with the rules for the construction of axiomatic formal theories, mathematical theories have a clearly presented and perceived structure comprising rigorously defined classes of symbols (an alphabet), rules for the construction of formulas (finite sequences of the elements of the alphabet), logical copulas and operators, auxiliary symbols, rules for constructing propositions out of formulas and logical copulas, and rules for the calculus of propositions. An ensemble of these elements, along with their interpretation, forms a mathematical theory, marked by exactness and rigorousness due to the unambiguous identity of its ingredients.

(e) Rejection of appeals to empirical experience as a means of controlling discourse. Mathematical propositions are necessary, but their necessity does not stem from direct empirical experience. The latter assumed, as we have pointed out. outsize proportions in the methodological conceptions of mathematics during the Modern Times. "...An empirical judgement," wrote Kant, "never exhibits strict and absolute, but only assumed and comparative universality (by induction); therefore, the most we can say is,—so far as we have hitherto observed, there is no exception to this or that rule" (151, 2). "Since experiential knowledge cannot be universal and necessary, universal and necessary knowledge cannot be experiential,"—this dogma was the reason why mathematics was regarded as an a priori science.

Does a priori science exist at all? Dialectical materialists unequivocally answer this question in the negative. Firstly, as we have stressed already, the propositions of mathematics are empirical in origin and are ultimately verified by practice; in this sense, they cannot be a priori, i.e., generally independent of man. Secondly, the assertion that a given proposition is universal and apodictic is in itself problematic, being a hypothesis that is later either confirmed or rejected. Highly instructive in this respect is lack of consensus among mathematicians as regards such propositions as the axiom of choice, the tertium non datur, the abstraction of actual infinity, etc.

None of this can refute the proposition, however, that mathematics does not appeal to experience as a means of verification of discourse; the means used here are the tests of consistency, completeness, independence, and deducibility from the axioms. These non-empirical criteria for the evaluation of mathematical knowledge, so abstract in character, make it possible to avoid the familiar errors of the empirical criteria and means (experiment, observation, statistical processing of data, etc.) of evaluation of knowledge, which increases the exactness and strictness of mathematics.

(f) Maximal limitations of the intuitive basis. Intuitive unexplicated elements—which are, as a rule, the cause of potential theoretical defects—are most rigorously excluded from theories that impose the strictest limitations on the qualities of systems that are possible in the abstract (i.e., contain the most definite claims). The latter feature is a consequence of the traditional mathematical processes of axiomatisation and the use of special symbols (numbering the positions and other means of eliminating the polysemy and non-rigorousness of everyday language) as well as of natural (unspecified) means of information translation.

At the same time the exactness and strictness of mathematical knowledge must not be regarded as absolute attributes. As in any other system of knowledge, mathematics has areas with problematic and strictly unsubstantiated elements, like mental experiments with abstract objects, risky hypotheses, surmises, daring assumptions, etc., which may, given the right circumstances, evolve into strict mathematical theories.

If we were to list at least some of the basic causes for the non-universality of mathematical exactness and strictness, the result would be as follows.

(a) In his lectures on the development of mathematics, Klein expressed the view that two relatively isolated stages in the evolution of mathematical knowledge could be discerned. Stage one: the search for new methods and heuristic instruments of solving internal mathematical problems (the work of investigation and research), at which there is practically no room for substantiating the innovations. Stage two: the substantiation of innovations (organisational activity).

The source of the exactness and strictness of mathematical theories being the mode of their formal axiomatic organisation, which is only realised at relatively late stages of mathematical activity, mathematicians' activity at the early stages is marked by imprecision and lack of strictness (in the sense of the absence of formal axiomatic foundations).

Another factor determining the absence of exactness and strictness at the initial stages of mathematical activity is induction. Induction, which is mostly incomplete in mathematical, as well as any other, knowledge (in principle, any non-trivial induction is incomplete), acts as a significant source of problematicity. Just one example to illustrate this idea.

Pierre Fermat formulated the proposition that numbers in the form $(2^{2^n} + 1)$ are always prime numbers, having proved this for $n = 1, 2, 3, 4$. But this general proposition, formulated by Fermat on an inductive basis, was refuted by Leonhard Euler, who showed that for $n = 5$, the number $(2^{2^5} + 1)$ is not a prime number.

To eliminate the inaccuracies resulting from the employment of the inductive methods of cognition, propositions obtained by induction are reformulated on the axiomatic deductive basis.

(b) Not all the propositions or formulas of mathematics are solvable. The existence of unsolvable problems follows, as a general case, from Gödel's incompleteness theorem, which reads: For any consistent formal system S containing a minimum of arithmetic there exists a formally unsolvable proposition or formula A such that neither A nor $\sim A$ are deducible in S. The existence of unsolvable problems is also established by Gödel's theorem concerning relatively unsolvable problems, the corollary, which follows from it, of the existence of absolutely unsolvable problems, and the Skolem principle concerning the impossibility of a system of axioms which embraces everything that it desires to embrace.

In 1935, Alonzo Church cited an example of an unsolvable mass problem, and later, jointly with George Rosser, established the unsolvability of elementary mathematics. In 1947, Andrei Markov and Emile Post proved the unsolvability of the problem of identity for semigroups; in 1952, Pavel Novikov proved the unsolvability of the problem of the identity of groups; later he showed the unsolvability of the problem of isomorphism (in group theory). In 1958, Markov proved the unsolvability of the problem of the homeomorphism of polyhedrons. In 1970, Yuri Matiyasevich proved the unsolvability of Hilbert's tenth problem. These examples show that unsolvability can be of two types: the unsolvability of propositions (a consequence of Gödel's theory of incompleteness) and algorithmic unsolvability (Hilbert's tenth problem). Attempts to clarify the concept of algorithm (Church's theory of λ-conversion, Kleene's recursive functions, Post's finite combinatorial processes, etc.) resulted in the general conclusion that an algorithm for the solution of the problem of solvability is impossible.

(c) Despite its explicit formal axiomatic organisation, the structure of mathematical knowledge includes a set of ideas that are, epistemologically speaking, intuitive, implicit, and not amenable to rigid deductive logical fixation and substantiation. To explain this proposition, let us merely cite a passage from Alonzo Church: "...any foundation of ... mathematics ... is in a certain fashion circular. That is, there always remain presuppositions which must be accepted on faith or intuition without being themselves founded" (127, 184). It follows that, since the definition of formal mathematics requires intuitive mathematics, mathematics as a whole is not exact in any absolute sense.

The most fundamental reasons for appealing to intuitive notions in mathematics are as follows.

(1) Its incomplete formalisability (cf. Gödel's limiting theorems).

(2) The axiomatic character of organisation; axioms are accepted without proof, largely on an intuitive and psychological basis. Characteristic in this respect is the evaluation of the axiom of choice by different mathematicians. Some of them (as e.g. David Hilbert) call it the premise of any mathematical discourse. Others, like Henri Borel, see it as a source of defects in mathematics. On the one hand, the axiom of choice is widely used in analysis, algebra and topology, in proofs of fully obvious theorems. On the other hand, it is a source of non-constructive proofs; besides, it is employed in proving propositions that are far from obvious (the Banach—Tarski paradox).

There are also difficulties in the substantiation of the axiom of substitution, and some others.

3) The use of non-rigorous concepts in the absence of any possibility of clarifying them. One of the causes for the antinomies of set theory is, according to some specialists, the use of Cantor's extremely vague definition of the set concept. Cantor defined a set as any multitude thought of as a unity. We shall not try to find faults with this formula or criticise the possibility and fruitfulness of interpreting a set as a "multitude thought of as a unity". It is hard, though, to get rid of the question: How is a multitude to be united, properly speaking, in a single whole? There is no precise answer to this question. At present, sets are specified either by listing their elements or by pointing to a well-formed predicate which constitutes a set on the basis of associating corresponding objects. It is clear, however, that a set is something different from a list of objects corresponding to a predicate. The nature of this difference has proved elusive so far.

There are also considerable difficulties in the use of the

entirely unexplicated concept of infinity, which is central to mathematics. It underlies the algorithmic mean-value theorems, the Taylor theorems, existence theorems, and so on. There is no proof, however, that its systematic and practically uncontrolled use does not stimulate paradoxes (in set theory). The concept of infinity illustrates Gottlob Frege's idea that there are signs in science (in language in general) which have no precise meaning although they express a certain sense. Such signs (as, e.g., "a three-headed man") are neither true nor false. They are epistemologically empty, so to speak, as we know practically nothing of their referents.

Frege insisted on the elimination of such signs from science, as in his view all well-formed signs had to denotate something.

In this respect, such concepts as infinity or the set of all sets are signs with unclear meanings. That may be the reason why it is not quite clear whether they should be excluded (as they are in intuitionism and the theory of types) or not, and if they should, what must be our attitude to them?

(d) Mathematics widely uses the apparatus of non-predicative definitions which include self-referential concepts and are based on the relation of self-application. Thus all the paradoxes of set theory are due, according to Abraham Fraenkel and Yehoshua Bar-Hillel, to the fact that "in all of them the crucial entity is defined, or characterised, with the help of a totality to which it belongs itself" (139, 54). Although it is clear that non-predicative definitions are not always destructive, it has proved impossible to find out in which particular cases they are destructive. The available suggestions for the choice of a criterion "to separate the lambs from the goats", such as Heinrich Behmann's suggestion for using in discourse only those terms whose ultimate eliminability can be proved, are not widely accepted in science..

As for the theory of types, in which non-predicative classes are excluded, it simply proves impossible to present an essential part of higher mathematics in its framework. It is thus practically not feasible to avoid non-predicative definitions, although the admission of non-predicative classes entails uncertainty about consistency.

Taking all this into account, one can accept John von Neumann's insistence on the relativity of the categories of exactness and rigour: "It is hardly possible to believe in the existence of an absolute, immutable concept of mathematical rigour, dissociated from all human experience" (174, 6).

The adoption of consistency as the central criterion of

scientificity. A mathematical (axiomatic) system is regarded as consistent if, for any proposition A, propositions A and $\sim A$ are not simultaneously provable in this system. Simultaneous demonstrability of propositions A and $\sim A$ in a certain system is evidence of its meaninglessness.

A discussion of consistency as a central criterion of the scientificity of mathematical knowledge cannot ignore this question: What should the attitude be to inconsistent mathematical theories? The apparently natural view that an inconsistent theory is unacceptable is not, however, either unambiguous or exhaustive. We refer here above all to the evaluation of set theory, the most fundamental of mathematical theories; it forms the foundation of mathematics, yet it is not consistent (it is not free from paradoxes). How are we to handle a situation, then, in which mathematicians recognise the exceptional role of the criterion of consistency yet show tolerance towards the inconsistencies of set theory which forms the foundation of mathematics?

There is no generally accepted clear answer to this question. In our view, a way out of this situation can be found on the following two approaches: (a) one may reject the traditional methodological values in mathematics (which will be tantamount to a rejection of the consistency criterion as the central criterion of mathematical knowledge), leaving aside the problem of substantiation of mathematics on a non-paradoxical basis; (b) one may retain the traditional methodological values in mathematics (which is tantamount to accepting the consistency criterion as central to mathematical knowledge), simultaneously building a suitable non-paradoxical foundation for it. Let us consider the prospects for these two approaches.

Should we link up the concept of scientificity of mathematics with rigour and consistency? It is a fact that in all other sciences we are content with much weaker demands, so it may be worthwhile to modify the traditional conception of scientificity criteria in mathematics. This is not just an abstract possibility—this approach is now gaining ground in mathematical science. There is a tendency to introduce new methodological values, or criteria of scientificity, into mathematics, such as practical utility, heuristic effectiveness, cognitive usefulness, etc. In their analysis of the crisis in set theory, B. Pyatnitsyn and V. Porus point out, for instance: "We believe that the paradoxes discovered in set theory have not become the 'destroyers' of this theory (and of the whole of mathematics) because, firstly, it cannot be squeezed completely into the logical dimension of axiological characteristics (that is to say, consistency is not here, in fact, an absolute value...), and secondly, and probably more importantly, a protective role is played here by the cloud itself of values with which set theory is shrouded in the space of evaluations or, in other words, in the consciousness of practical mathematicians" (79, 77).

On the whole, this approach, as illustrated in the passage cited here, seems to be an *ad hoc* orientation. It is a fact that set-theoretical paradoxes violating the demand of consistency have brought no satisfaction to anyone. The actual use of set theory in mathematics is not a result of a reappraisal of the traditional values in this science, neither is it a result of the action of the cloud of additional values that is said to shroud set theory. The causes of this situation are simply these: (a) the paradoxes of set theory are not elements of its central or useful part, as von Neumann puts it; (b) a better system that would be equivalent to set theory does not exist; (c) at present, there is no acceptable programme for an allround substantiation of mathematics.

Underestimation of the paradoxes in set theory, reflected, e.g., in their description as sophisticated artificial constructs, does not appear to be convincing. Fraenkel's and Bar-Hillel's evaluation of the situation is rather emotional; besides, it does not follow from their line of investigation; cf. their statement: "many a scientist wishes that his field were in as 'critical' a state as mathematics, and few are the mathematicians who are really depressed by the existing uncertainties in the foundations" (139, 270).

One can easily find statements directly opposite to this in tone. Suffice it to recall Hilbert's view: "...The position in which we find ourselves as regards paradoxes is, in the long run, unbearable. Consider: in mathematics, this model of certainty and truth, conceptual structures and conclusions, as everyone learns, teaches and uses them, lead to absurdities. And where are certainty and truth to be found when even mathematical thinking fails us?" (144, 274).

Thus the appeal to certain secondary methodological values in mathematics, which does not cancel the need for its substantiation in accordance with the requirement of consistency, is in our view a sophisticated ruse intended as a means of temporarily relieving the acuteness of the situation engendered by the paradoxes in set theory. Inasmuch as an inconsistent theory brings no satisfaction, and no reappraisal of values can dispel the dissatisfaction with an inconsistent theory, we see more promise in a different approach—one based on the desire to find a foundation for mathematics in accordance with the requirement of consistency.

For lack of space, we cannot go into a detailed methodological evaluation of the familiar trends in the foundation of mathematics. We concur with Mostowski's view that the philosophical goals of the three schools in the foundations of mathematics (logicism, formalism, and intuitionism) have not been attained, and that we are now as far from a complete understanding of mathematics as were the founders of these schools (173); and also with Kleene's statement that the paradoxes "leave us the problem of refounding set theory on a drastically altered basis, the details of which are not fully implicit in the suggestions' (153, 40).

The problem of the foundations of mathematics is a problem in the search for reliability, rigour, and exactness, that is, in the final analysis, a search for all that is associated with scientificity. The problem of foundations cannot be solved without clarifying the following problems connected with it (and, most likely, it has not been solved because they have not been clarified): (a) the causes of the paradoxes; (b) the status of mathematical objects (the solution of the nominalism—realism—conceptualism trilemma); (c) the problem of existence; (d) the nature of mathematical statements; (e) the subject-matter of mathematics.

Progress in the elaboration and foundations of set theory can only be achieved through deepening general notions of mathematical reality. Indeed, since paradoxes are questions that cannot be answered as long as there is no general picture and no clear understanding of connections in a given domain, elimination of paradoxes from set theory requires that such a level of understanding should be achieved, which in this case can only be done through a deeper perception of the essence of mathematical reality and the nature of mathematical activity as a whole. At present we do not know enough about this nature, about those of its elements that lead to paradoxes, etc. That is why, however carefully we evaluate the available programmes in the foundations of mathematics, we do not know how these elements are to be eliminated by rebuiling set theory. It is therefore true that "no unique and universally accepted way of reconstructing mathematics exists or is in view, and in this sense the foundational crisis is still in force" (139, 270).

The features of mathematics cursorily outlined in the above

reflect general rather than particular properties inherent in systems of mathematical knowledge as such.

3.2. NATURAL SCIENCE

In its most direct usage, the expression "natural science" denotes a broad cognitive domain oriented towards the study of primary objectively existing nature in its immediate givenness. Seen as a generic form of knowledge, natural science is heterogeneous and multidimensional. It embraces a great number of disciplines, planes and structures united by their orientation towards theoretical assimilation of a general substantive basis termed *materia naturata.*

Natural-scientific cognition of matter is focused on "operations of its allround analysis, description and explanation of forms and mechanisms, structural decomposition, establishment of elements, qualitative and quantitative values, limits, measures, optimal and critical conditions of existence, interaction with the environment, and so on" (50, 28).

From the typological standpoint, natural science consists of empirical-phenomenalist (descriptive) and theoretico-essentialist (explanatory) theories.

Empirico-phenomenalist natural-scientific theories are ensembles of qualitative phenomenological systems of knowledge; predominant among them are (a) empirical descriptions recording in language terms data obtained at the empirical level of the study of the object through measurement, observation, analysis and choice of facts, visual registration, primary classification and systematisation, various types of experimentation, etc. (cf. the description of the phenomenon of inheriting of contrasting features in hybridisation before Georg Mendel formulated the appropriate laws), and (b) empirical laws, dependences, regularities obtained as inductive generalisations of experimental data (the Mendel laws before the chromosome theory of heredity asserted itself in science). The mechanism of the transformation of descriptive, cumulative knowledge (i.e., of practically all natural-scientific knowledge at the initial stage of its development) into explanatory and systematic knowledge is metrisation and the corresponding logico-semantic substantiation.

Explanatory theories are an ensemble of logically organised systems of knowledge; predominant here are (a) theoretical explanations conceptually reconstructing the data obtained at the theoretical level of the study of the object through interpretation, idealisation, mental (conceptual) experiments, abstract modelling, etc. (cf. the Mendel laws obtained at the representative level as consequences of the chromosome theory of heredity); the bulk of the components (propositions) of theoretico-essentialist natural-scientific knowledge is deduced from fundamental propositions—postulates and axioms; (b) exact quantitatively detailed results (the distributions of the contrasting features in the first and the subsequent generations of hybrids that were quantitatively detailed by Mendel and therefore assumed the status of numerical laws).

Among explanatory theories, hypothetico-deductive and axiomatic theories can be justifiably distinguished.

Hypothetico-deductive natural theories are based on the application of the hypothetico-deductive method which occupies an intermediate position between the properly empirical and axiomatico-deductive methods of research, being founded on logical derivation of consequences from hypotheses and their subsequent factual verification. Classical mechanics was constructed in accordance with that method. Thus Newton first introduces the fundamental concepts of that science and then its laws and propositions subject to verification. A special case of the hypothetico-deductive method is the method of the principles of theory construction. A theory figures here as an ensemble of experimentally verified propositions logically derived as consequences from basic principles. The method of principles underlies thermodynamics, the special theory of relativity, the general theory of relativity, etc.

Theories subjected to strict logical reconstruction and elaboration are called axiomatic. The orientation towards axiomatisation and formalisation of natural-scientific knowledge is quite understandable in view of the fundamental difference between a formalised meaningful theory and the original meaningful theory—the fact that the propositions of a formal theory are independent from interpretation. Since the organisation of natural-scientific knowledge on the principles of axiomatisation and formalisation assumes the specification of groups of axioms precisely stating the logical, mathematical, and properly scientific foundations of a theory, which is only possible in an advanced theoretical science, it is clear why many natural-scientific theories remain unaxiomatised and unformalised. In biology, for instance, we only come across isolated attempts at axiomatisation, such as Woodger's axiomatic version of Mendel's genetics using the language of *Principia Mathematica* (195). A gratifying exception in this respect is physical knowledge, which attained the level of advanced theoretisation earlier than the other sciences. Constantine Caratheodory, John von Neumann, Arthur Wightman and others undertook, at different times and with varying success, formal-axiomatic substantiation of its isolated (though extensive) fragments. Relevant to the axiomatisation of physics as the basis of natural science is Hilbert's sixth problem.

It is justifiable to isolate in the framework of natural science a large area of knowledge that is not explicitly connected to the types of natural-scientific theories specified above. Of this nature are, for instance, many branches of geology, tectonics, palaeontology, biology, soil science, geography, climatology, cosmology, etc., occupying, as it were, an intermediate position between descriptive and explanatory theories, as they use the method of historical reconstruction based on combined application of empirical and theoretical studies.

Apart from logical and philosophical premises, the structure of a natural-scientific theory includes its own foundations, i.e., the basic principles and postulates of the theory which determine all its propositions. The actual body of the science consists of facts, laws, and propositions concerning facts and laws, which form a superstructure built on the theory's own foundations.

A fact is not necessarily an empirical statement containing observation data. Epistemologically, a fact may be a statement pertaining either to empirical or to theoretical knowledge, if its truth has been established with certainty. A fact is, above all, an element of knowledge that plays a definite role and has a certain function in the system of knowledge. It would be hard to exaggerate the cognitive role of facts in defining the domain of a theory, in substantiating its propositions, in establishing the empirical meanings of theoretical models, etc. In all these cases, as Engels points out, "one must proceed from the given *facts*" (59a, 47).

As distinct from the sphere of everyday knowledge, the establishment of a natural-scientific fact is connected with seriation, that is, with statistical averaging of a series of experiments. Natural-scientific facts are therefore always average statistical summaries, the frequencies of occurrence of a definite feature or characteristic of an object.

A natural-scientific fact—an epistemologically complex conceptualised structure—is set against the naked empirical fact, which is a real phenomenon under concrete conditions. A naked empirical fact is an objective event recorded by the sense organs and considered as uninterpreted and not transformed subjectively. An illustration of a naked empirical fact would be the perception of the deflection of the needle of a dial noted down in the language of an observation record: "... At the moment of time to the needle of the ammeter moved 10 points to the right of zero". Any attempts to give an interpretation of this phenomenon, however, will necessitate its inclusion in the context of a theory explicating this fact. Strict natural-scientific facts should be distinguished from distortions of reality on the basis of a biased theory, such as facts that are alleged to confirm inheriting acquired traits, etc.

A law is a central proposition of a theory recording essential, necessary, recurring, or universal connections of the objective world. A law is an expression of the form

$$\forall(x) \quad (P\ (x) \rightarrow Q\ (x))$$

which is not an abbreviated notation for an infinite conjunction of propositions (this would be in keeping with the inductivist interpretation of the generality quantifier), as it does not state something about separate elements of the class but about a necessary and universal property of these elements. Close formal-logical investigation of the nature of universal implicative propositions has brought to light paradoxes known as the paradoxes of material implication. The nature of these paradoxes follows from the formal-logical orientation of the analysis of the laws of science, which are interpreted in terms of their logical form regardless of the content they express.

From the logical standpoint a law is a conditional relation in implicative notation. The latter, however, is also the form of expression of factually true general propositions of the type "all ravens are black", which do not have the status of laws. The problem arises of isolating the subset of laws in the set of all universal implicative propositions (the so-called Lewis problem). The gist of the Lewis problem lies in the question: can an implicative relation identical with the meaningful relation between propositions connected by implication be constructed in the system of propositional logic? Lewis himself came to the well-founded conviction (161) that such a relation cannot be constructed in the system of classical logic.

Hence the interest of non-classical logics intended to solve this problem—systems with strict implication (Lewis), strong implication (Ackermann), with the Nelson implication, causal implication (A.W. Burks), etc. But none of these systems contain a solution of the problem. Although it was shown later that the criterion differentiating laws (nomological propositions) from accidental universal propositions in implicative form is the deducibility from the former of conditional counterfactual propositions demonstrating the existence of true necessary relations, the difficulties should be pointed out of establishing nomology by reformulating universal implicative propositions as counterfactual ones. The difficulty lies in the fact that there is no formal logical procedure for substantiating counterfactual propositions; their truth is established factually, through analysis of the satisfiability of the material relations ensuring the nomological character of the connection between consequent and antecedent. That is precisely where the difficulty lies, for carrying out this type of analysis is not a trivial task at all.

Propositions concerning facts and laws form the semantic corpus of a theory. A natural-scientific theory is a set of propositions ordered by the relation of deducibility of varying degrees of rigorousness specified for it, or, in more formal notation,

$$T\ (C, A, L) = \{P_1 \ldots P_n\},$$

where T is a natural-scientific theory; C, basic or derived constructs; A, a set of axioms; L, derivation rules; P, the propositions of the theory. P includes propositions about individual facts, empirical dependences and theoretical laws.

Propositions about individual facts are formulated in terms of the language of direct observation (i.e., they do not include quantifiers) at the empirical level of research. Propositions about empirical dependences are written down in the language of empirical constructs and include terms that do not pertain to direct observation but correspond to the empirical level of research (that is to say, they do not assume explanation). Propositions about theoretical laws are expressed in the language of theoretical constructs, include terms of the two above-mentioned types of sentences, contain propositions with quantifiers, and correspond to the theoretical level of research (105). These three types of propositions are connected by reduction rules specifying the modes of transition from the language of observation (propositions about the readings of a measuring device) to the language of empirical and theoretical constructs (e.g., statements about objects acting on the measuring devices, etc.).

The formula for a natural-scientific theory is as follows:

$$\Omega = \langle Fct, Lw, Cnst, Int, Abstr\ mdl, Frml, L\rangle$$

Fct is a non-empty set of facts corresponding to reality.

Lw is a set of laws; in exact natural-scientific theories establishing and predicting the properties of objects on the basis of some initial data (cf. the behaviour of mechanical systems predicted on the basis of the knowledge of forces and of initial conditions), this set is not empty; in inexact fields of natural science concerned with accumulation of facts and operating as a rule with qualitative correlations, this set is empty.

Cnst is a set of constants; in the case of many theories this set is empty.

Int is a set of natural, empirical and semantic interpretations ensuring the identification of *Abstr mdl* and *Frml* of natural-scientific theories with the real prototypes.

Abstr mdl is a set of abstract models, semantic assumptions, and auxiliary constructs accepted in a theory.

Frml is a set of formalisms acting as instruments of thinking, language, and quantitativisation of propositions of natural science, which permit it to operate with quantitatively detailed connections of reality, i.e., to formulate dependences of the type $x = ky$.

L is a set of logical assumptions and derivation rules accepted in a theory.

Substantive specification of these sets requires the introduction of the concept of *Srt*—a non-empty set whose elements are called sorts of ontologically individualised objects. Thus if each of the sets listed here is correlated with variable sorts $\pi \varepsilon Srt$, the theories will be object-specified.

Considering that natural science is a set of substantive theories (*Ts*) described by Ω, and taking into account the significance of the apparatus of logic (*L*) (the instrument of substantiation) and mathematics (*M*) (the instrument of raciotination in natural science), its structure can be expressed by the formula $<\Sigma\ Ts;\ M;\ L>$.

The epistemological specificity of natural science consists in the following.

There exists a direct (rigid) correlation between natural science and a definite fragment of reality; this feature distinguishes natural science from mathematics and makes it explicitly ontologically specified. Being a reduced reproduction of reality, theory studies material relations between objects in fixed domains, which determine the qualitative traits of both the separate elements and of their entire internal structure. The dependence between the conceptual basis (B_c) of the theory and its empirical basis (B_e) can be expressed, for ease of visual perception, by the formula $B_c \leftrightarrow B_e$ (62, 110), where the mechanism of links between the bases (expressed by $\leftrightarrow$) and B_e (neither of which figure in mathematics) epistemologically specify the essence of natural science.

B_c, formed by *Lw, Cnst, Int, Abstr mdl, Frml, L,* acts as a categorial, semantic and organisational schema, the framework of theory, with which all the numerous particulars are linked up. The creation, elaboration and construction of B_c are marked by freedom (the considerations of convenience, simplicity, and temperament affecting the choice of *Frml, Int, Abstr mdl*) but not by arbitrariness. This is explained by the orientation of B_c towards reflecting the objective connections of reality, from which it follows that there exists a certain fundamental limit of the subject's freedom in the framework of B_c, namely, the objective logic of the functioning and development of the real object. This logic, permitting freedom, excludes arbitrariness. The object domain does not, of course, impose any fixed definitions on B_c, but at the same time it rules out arbitrary constructions. Thus, it was precisely the pressure of the object logic that led to the elimination of ether from physics. That is probably the reason why the desire to retain ether at any cost in the B_c of physical theories (148) appears so artificial and unconstructive.

B_c is correlated with B_e as a whole, as a systemic mutually coordinated structure; many components of B_c facilitating mediated modelling of reality are of intermediate, subordinate character, and are not by themselves projected onto empirical facts; they have no analogues in reality. Of this nature is, e.g., the Ψ—function in quantum mechanics, of which only the square of the modulus has an empirical meaning.

B_e, formed by facts, acts as an external source of the content of theories, as well as an instrument of their substantiation. On the dynamic plane, the theory is realised as successive accumulation and systematisation of data (observation, establishment of empirical dependences), their theoretisation, derivation of empirically discoverable consequences from the systems thus obtained, final justification of theories, and their implementation in practice. The connection between B_c and B_e is not immediate. That means that, strictly speaking, the content of B_c does not correspond to the relations of the real world but of an idealised one, a world of secondary conceptual reality which represents the analysed objects of primary or objective reality. Although objective and scientific realities do not coincide (the opposite assertion leads to naive realism, which is a trivialisation of the problem), they are linked by necessary ties (rejection of these ties is fraught with the heuristic blind alleys of Platonism), and these ties are due to the special role of B_e. First, all components of B_c have an empirical genealogy, which may sometimes be difficult to reconstruct but is always real. Second, B_e specifies the vector of the progress of B_c towards a better balance between B_c and B_e. Third, B_e is "the alpha and the omega of all our knowledge of reality" (136, 271). B_c "is made up of concepts, fundamental laws which are supposed to be valid for those concepts and conclusions to be reached by logical deduction. It is these conclusions which must correspond with our separate experiences; in any theoretical treatise their logical deduction occupies almost the whole book" (ibid.). Fourth, being relatively independent from B_c, B_e functions as an informative and critical instance, preventing complete and final identification of theory and reality, and the substantivisation of conceptual models.

Any theory is a theoretisation of facts; theories are modified and discarded, while facts remain and are preserved. The stability of facts in theories succeeding one another is the ontological basis of their commensurability and of establishing the degree of progressiveness of the theoretical shifts.

"$\leftrightarrow$" is composed of the rules of correspondence, or reduction rules, and algorithms of links between theoretical constructs and the world picture (through models, through interpretation) which is projected onto B_e, bearing in mind that B_e is not pure reality as such but conceptually and operationally mediated reality specified in terms of experiment, measurement, and human experience.

Correspondence rules are intended for the verification of the empirical content of theoretical terms, ensuring, through a system of operational definitions, the transition from models, interpretations and formalisms to facts, permitting comparison of the consequences derived from B_c with interpreted experimental results and thereby an experimental substantiation—confirmation or rejection—of a theory. Correspondence algorithms realise the meaningful, communicative, and explanatory processes in natural science, determining the theoretical load or "opaqueness" of B_e and mutual connections between models and interpretations on the one hand, and facts on the other. Thus Maxwell's electrodynamics interprets electromagnetic processes as time variation of electric or magnetic intensity and density of current at a point, whereas the cognate electrodynamic theory of Ampère and Weber interprets the same processes as time variation of the states of ether—the conducting medium of electromagnetic phenomena.

A detailed study of the cognitive role of "$\leftrightarrow$" compels a consideration of the mechanism of links between B_c and B_e in a more general form, which requires further clarification of the epistemological specificity of natural science.

There is no direct logical bridge between B_e and B_c, which is tantamount to the impossibility of direct deduction of B_c from B_e or of reduction of B_c to B_e. Rejecting the possibility of direct deductive-reductive relations between B_c and B_e, we stress the creative essence of B_c, which emerges in the process of synthetic productive activity that has nothing in common with direct generalisation of B_e. Any attempt at direct logical derivation of the basic concepts and laws of natural science from elementary experience is doomed to failure. At the same time the autonomy of B_c and B_e is not absolute, as they are connected by mediated deductive-reductive links.

The proposition that B_c follows from B_e implies empirical origin of B_c rather than its deductive derivation.

Empirical genealogy of B_c must not be interpreted in the spirit of primitive inductivism, as shown by the dynamics of the B_c of a natural-scientific theory at the stage of advanced theoretisation. The unfolding of the B_c of a theory through generative logico-mathematical procedures, operations of expanding synthesis, and meaningful operations with objects never results from direct inductive generalisation of B_e but, being empirical in origin, is directed towards B_e.

The building and formation of the body of B_c through generative logico-mathematical procedures and meaningful operations with objects consists in obtaining within a given theory of results by means of deduction, as logical consequences of manipulation of theoretical objects. First, a definite system or object of operation is specified, to which logical, mathematical and semantic operations are applied; as these operations are performed in a definite sequence on definite components of the object of operation, systems of operations are formed, leading to certain results. Further unfolding of B_c assumes the formation of ever new generative procedures and operations, each of which goes further away from the basis of B_c and, relying on previous results, produces new ones. In this way, a closed-in tree-like theoretical structure takes shape.

However, deductive self-branching of B_c is not unlimited. Owing to the pressure of B_e, the need arises, sooner or later, for expanding the substantive foundation—the basic principles of B_c, for feeding in fresh ideas on the procedures and operations generating it.

The obligatory nature of the generative logico-mathematical procedures and meaningful operations on idealised objects within B_c for the formation of its body discredits the proposition that B_c is derivable from B_e through inductive generalisation: there is no logical path from facts to theory. The fact that expanding synthesis operations are obligatory for potential modifications of B_c determines its links with B_e.

The property of reducibility in B_c and B_e does not imply direct reducibility of B_c to B_e but only the justifiability of B_c in terms of B_e. If B_c is not connected in any way with B_e and cannot, even potentially, be projected onto it, it is devoid of empirical foundation (cannot be experimentally substantiated). In a situation like this, doubts creep in about the seriousness, acceptability and reliability of B_c, which is epistemologically relegated to the class of arbitrary constructs and natural-philosophical speculations.

B_c is not reducible to B_e, because its infrastructure—the theoretical principles and idealisations—are ideal and not derived from experience. These are, as Einstein pointed out, "free inventions of the human intellect, which cannot be justified either by the nature of that intellect or in any other fashion *a priori*" (137, 272). At the same time the autonomous status of the logical structure of B_c does not dissociate it from B_e: "The empirical contents and their mutual relations must find their representation in the conclusions of the theory. In the possibility of such a representation lie the sole value and justification of the whole system" (ibid.). In other words, the impossibility of projecting B_c onto B_e means that B_c has no natural-scientific meaning. The freedom of the scientist's activity in the framework of B_c is not absolute: "...It is not in any way similar to the liberty of a writer of fiction. Rather, it is similar to that of a man engaged in solving a well-designed word puzzle. He may, it is true, propose any word as the solution; but there is only *one* word which really solves the puzzle in all its parts" (137, 294).

Rejection of direct reducibility of B_c to B_e makes it possible to avoid the dogmatism of naive realism. Recognising that B_c is mediatedly reducible to B_e, we can regard B_e as a criterion of the truth of B_c, thus avoiding the relativism of conventionalism.

Thus the links between B_c and B_e are multidimensional and dialectical. Underestimation of or neglect for this fact entails various methodological hypertrophies. A correct interpretation of the problem is only possible on the assumption that any B_c, however abstract it may be, is empirical in genesis and has an experiential-practical substantiation. If theories were not generalisations of experience, they would be unable to predict anything and neither could they be confirmed. But that is not the whole point. If theories were only generalisations of experience, they would never contradict the latter.

An indication of the absence of one-to-one dependence of B_c on B_e is the phenomenon of "equivalent formulations". Thus quantum mechanics can be formulated in terms of matrices (Heisenberg) and waves (Schrödinger), but the principal concepts and correlations of these formulations remain invariant with respect to the formal apparatus employed. The content of classical mechanics can be expressed to the formulations of Newton, Lagrange, Jacobi—Hamilton, etc.

Numerous axiomatic formulations of various theories indicate that their content can be equally well expressed in different ways. Generally speaking, true laws of natural science permit, as Richard Feynman points out, a great number of different formulations (138, 55), which refutes all uncritical postulates concerning the existence of a one-to-one connection between B_c and B_e.

Mathematics is used as an instrument (the logic of thinking) and the language of cognitive activity. The fact of striking effectiveness of mathematics in natural science has always intrigued scientists and methodologists. However, lack of clarity in this field gave rise to a sort of confusion. The effectiveness of mathematics in natural science was often described as incomprehensible (193) or unknowable (121). We do not share these assessments and moods; pessimism has never been constructive. The deep causes of the effectiveness of mathematics in natural science appear to us quite knowable and intelligible. Omitting the details of various opinions, one may summarise these causes as follows.

(a) Unconfined by the limits of an object domain, mathematics possesses great heuristic and research possibilities. The conditions under which a mathematician works are comparable to those of a science-fiction writer. Indeed, what are the causes of the striking realism and exceptional perspicacity of many predictions of science fiction? This ability for clairvoyance is explained by freedom of activity; there are no difficulties of practical realisation of ideas, the creative modelling of more or less verisimilar situations is restricted by one factor only—the requirement of internal consistency, inner harmony, and coherence of the process and the result of construction. The same is observed in mathematics. Analysing an object in its pure form as a kind of logical possibility, a mathematician anticipates its potential substantive meaningful study in the corresponding theories. Thus, the Lobachevsky geometry emerged as the result of studying the possibility of constructing a geometry on the basis of the axioms of Euclidian geometry in which the parallel axiom was modified. The new geometry constructed by Lobachevsky was an "imaginary" abstract mathematical structure. However, it was later used in the special theory of relativity (the space of velocities of relativistic mechanics is a Lobachevsky space), in cosmology (in Fridman's open models, the spatial cross-section in the frames of reference accompanying matter is described by the Lobachevsky geometry), and so on.

What are the causes of this? What is the explanation of the close link between mathematical structures and experimental phenomena, as illustrated here?

The strength of mathematics is in the abstract universal study of its object—a study that can only be formal. In considering the logical possibility of something, a mathematician analyses the object in a maximally general form. The result of the analysis is substantively undetailed structures satisfying the criterion of consistency. Being consistent, mathematical structures prove to be an inexhaustible source of natural-scientific interpretations; they can be specified in various ways in terms of objective ontology depending on the needs of research.

Wherein lies the magic strength of mathematical structures, properly speaking? In the fact that they contain the truth (as comes to light *post festum*) before objective knowledge is reflected in the formally expressed truth. Thus mathematics lays in a store of truths, as it were, for they are perceived as truths only some time later (after a suitable interpretation is found). Natural-scientific cognition consists of two planes: the general plane (quantitative-formal, or mathematical), and the particular plane (qualitative-substantive, or natural-scientific). They are not synchronised in time, and that explains the gap between the moment of generation of mathematical structures, which can be used as the basis of a natural-scientific theory, and the moment of completion of the latter, at which interpretations of mathematical structures are found. There is nothing mystical or unintelligible about this, of course, as shown by the example of the Lobachevsky geometry.

Let us now consider the second question, formulated above, about the causes of links between mathematical structures and natural-scientific phenomena. On this plane, the effectiveness of mathematics in natural science is explained by the fact that the propositions of both mathematics and of natural science can be quantitatively detailed, they can be associated in one way or another with the category of number or magnitude; therefore, if mathematical structures are interpreted as quantitative correlations between magnitudes, with which certain real properties are associated, they acquire referents and become applicable to reality. The mechanism of translation of mathematical structures into the language of natural-scientific experimental phenomena is, as we know, correspondence rules, which include operation definitions. Let us stress that there are two planes again in the study of number—mathematical and natural-scientific. Mathematics takes a formal view of number—either in terms of the Zermelo axioms or in terms of the von Neumann axioms. Natural science takes a substantive view of number—in terms of operational definitions. These planes, however, can be superimposed in the natural-scientific interpretation of numbers as measures, magnitudes, measurements, etc. It is therefore easy to see that the secret of extraordinary effectiveness of mathematics in natural science lies in the identification of formalisms and magnitudes, in the units of measurement. Figuratively speaking, these are all nails fastening together mathematical structures and natural-scientific phenomena. In the case of the Lobachevsky geometry, these nails were the quantitatively detailed propositions of the theory of negative curvature space metrics which, having originated on the abstract mathematical, substance-free plane, was later applied in the special and general theories of relativity and in other theories as it was transposed onto the ontologically specified natural-scientific plane.

(b) Mathematical study of an object is inseparable from the translation of a given problem from the intuitive-meaningful into the formal-axiomatic language, which makes non-rigorous qualitative notions strict and exact, thereby expanding the heuristic horizons of research. The view that mathematics is like a millstone, grinding only that which is poured in, is not too profound. Mathematics is a creative science, and its creative nature is manifested in natural science in the use of ideas and constructs elaborated in mathematics—including the following.

(1) Formal constructs, which literally act as archetypes of future substantive meaningful natural-scientific constructs and theories. Thus, spinor concepts were first developed by Cartan as purely mathematical ones. Later, however, Dirac found them to be "field-magnitudes of a new sort, whose simplest equations enable one to a large extent to deduce the properties of the electron" (136, 274). An abstract mathematical schema became a concrete natural-scientific notion. Examples of this type are innumerable. Early in this century, James Jeans recommended to exclude group theory from the curriculum of Princeton University on the assumption that it would never be applied in physics; in 1961, Gell-Mann predicted, on the basis of this theory, the existence of the then unknown particle omega-minus, discovered in 1964 by William Fowler and Nicolas Samios. The theory of complex variable functions worked out by Augustin Cauchy and Georg Riemann was introduced in the theory of electric circuits. Functional analysis developed by Hilbert, and matrix theory formulated by Cauchy and Charles Hermite, are successfully used in quantum mechanics. The mathematical theory of canonic systems of differential equations developed by Gibbs was an important factor in the advances in statistical mechanics. In all these and numerous similar problems calculations, far from replacing ideas, stimulate them—in contravention of the Dirichlet formula. Let us consider this point in greater detail.

Mathematics produces ontologically unspecified structures; natural science realises only those of these structures that have meaning in its own framework; a revision of the natural-scientific "meaninglessness" of some mathematical structures entails, as a rule, the penetration of new ideas in natural science. For example, when Dirac set himself the goal of formulating the equation for a spinning particle which would satisfy the requirement of relativist invariance, he took an equation with a double solution as his point of departure:

$$E > E_o = m_o c^2 \qquad (1)$$

and

$$E < -E = -m_o c^2 \qquad (2)$$

From the physical standpoint, (2) is meaningless, as meaningless as many mathematically meaningful roots of *n*th-degree equations. However, Dirac did not reject the possibility (we repeat, the physically meaningless possibility) of a negative solution; the search for an interpretation of this solution led to the idea of existence of the positron, which was predicted in 1931 and discovered in 1932.

(2) Ideas of harmoniously elegant relations conforming with the principles of symmetry. Historically realised programmes of mathematisation of natural science are connected with these ideas. Among them are the ideas of number (Pythagoras), regular polyhedrons (Plato), perfect geometrical figures (Eudoxus, Ptolemy), etc., expressing the idea of quantitative proportionality—ideas that made a noticeable impact on natural-scientific research. If we accept that an inalienable concomitant of mathematisation is the awareness of the applicability of mathematics to natural scientific phenomena, a common feature of all programmes of mathematisation of natural science will be recognition of the fact that nature is the realisation of the most elementary mathematically thinkable elements, and that it is possible to find, by using purely mathematical constructs, the concepts and the regular connections between them which provide a key to an understanding of the phenomena of nature. For instance, it followed from Faraday's experimental works that rot $H = 0$. Maxwell added the missing term $\frac{1}{c}\frac{\delta E}{\delta t}$ without any experimental substantiation whatever. What was that step prompted by? It is hard to establish now the actual facts of the matter, but Max Born's explanation appears convincing. In his view, that step was prompted by Maxwell's desire for achieving the mathematical ideal of perfection, harmony, and beauty (119) that was not attained in electrodynamism as elaborated by Faraday. It was this striving for a mathematical ideal that compelled Maxwell to make an arbitrary addition to the equation.

(3) Formal viewpoints, which restrict, in a sufficient degree, the infinite variety of possibilities.

A world without limitations on diversity would be entirely chaotic (William Ross Ashby). In science, the instrument for introducing order in the world is theory. Theories present a rough, schematic, and idealised picture of the world, seeing it in terms of a finite set of basic principles. In their turn, the basic principles, the images and the systematic connections between them can, as a rule, "be arrived at by the principle of looking for the mathematically simplest concepts and the links between them" (136, 275). The effectiveness of mathematics lies in this case in the small number of heuristic schemata acting as models of diverse phenomena. On the one hand, the number of mathematically possible elementary types of correlations between natural phenomena, and of elementary equations that are possible between them is limited; that is the basis for the application of mathematics as an instrument of intensive study of the world. On the other hand, the freedom of mathematics from any links with a concrete ontological domain permits the elaboration of substantively universal formalisms uniformly describing the properties of objects of diverse nature; that is the basis for using mathematics as an instrument of extensive cognition of the world.

(c) The language of mathematics, an extremely convenient instrument, optimises natural-scientific activity. Each theory is correlated with a mathematical language of its own. In classical mechanics, that mathematical language is the language of numbers and vectors; in relativistic mechanics, the language of four-dimensional vectors and tensors; in quantum mechanics, the language of operators, etc.

The changes in the mathematical language used, e.g., in physics, are a good indication of the stages in the growth of that science. The programme of classical mechanics, accepted in pre-twentieth century physics, was based on the assumption of reducing all physics to mechanics. But the apparatus of ordinary differential equations used in the latter could not describe thermal, electric, and other types of phenomena. For this reason Fourier suggested the more flexible apparatus of partial derivative equations. Experience showed, however, that apparatus was not universal either, as neither the special theory of relativity nor quantum mechanics could be formulated in terms of the differential-analytical approach. At present, the mathematical foundations of physics includes different components. After it was proved that it was impossible to reduce the content of physics to the content of mechanics (and correspondingly, to reduce the mathematical apparatus used in physics to ordinary differential equations), that foundation absorbed, in addition to the differential-analytical, the set-theoretical approach (in the special theory of relativity), as well as the differential-geometrical (in the general theory of relativity) and the functional-analytical (in quantum mechanics) approaches. Their promising synthesis underlies the programme for the construction of the physics of the future.

The freedom of choosing mathematical apparatus for corresponding theories is limited by the pressure of empirical facts, by the need to take into account the existence of the objective logic of the given domain; in the final analysis, it is this objective logic rather than the mathematical apparatus that determines the positive content of the theory.

Natural science is an association of experimental sciences connected with concrete fragments of reality; the choice of a certain type of mathematical apparatus must be preceded by a careful analysis of its adequacy in the sense of agreement with the content of experience reflecting the appropriate fragment of reality. For the natural scientist, of the greatest importance is the identifiability of the mathematical apparatus with certain magnitudes—only in this case can it perform the descriptive, generalising, codifying, normative, and other functions, only in this case can it assert something about objective reality.

A consistent mathematical apparatus may be unacceptable as an instrument of describing reality in one theory, yet it may prove quite acceptable in another. The general foundation for this fact is the assumption that consistent mathematical structures can be given substantive interpretation. As for the sources of fundamental applicability of mathematical apparatus to the description of reality, they lie in the empirical origin of mathematical structures. This last proposition is substantiated in the dialectical-materialist theory of reflection, which permits the development of the most adequate epistemological theory of science without defects or blind alleys.

(d) Specifying the principles of objective fixation of results in the form of the requirement that equations (formulations, laws) must be invariant in respect to groups of transformations, mathematics acts as a kind of guarantor of the objectiveness of natural-scientific knowledge. This will have to be explained. The point is that the equations of abstract mathematised natural science do not directly describe the behaviour of material objects. Being formulated with respect to idealised or constructive Reality, they describe the behaviour of abstract objects—mathematical points (in classical mechanics), point events (in the special theory of relativity), etc.—which have the status of models vis-à-vis their objective analogues. Quite clearly, the requirement of invariance of equations describing the behaviour of idealisations relative to groups of transformations cannot guarantee objectiveness in the sense of coincidence of natural-scientific knowledge with reality. Their truth can only be guaranteed by practice, by empirical verification, by experiment. And yet we can and must regard the requirement of invariance of the equations of a natural-scientific theory relative to groups of transformations as a guarantor of their objectiveness.

The principle of invariance of mathematical formulations of a natural-scientific theory relative to groups of transformations is, in the most precise sense, a characteristic of research activity. Imposing quite concrete demands on the latter, it constitutes general rules for operating with abstcract objects and specifies algorithms of objective fixation of results. "Objectiveness" in this case means anti-subjectiveness and universality of the mathematical formulations of theory, which are of course an indication of their regular, necessary, and therefore objective status. The requirement of the invariance of the equations of a theory relative to groups of transformations is implemented in the following imperatives: in the framework of any theory, results must be independent from the specific features of their description in different frames of reference from the numerical magnitudes of parameters; the form of propositions must be independent from the units of measurement; and equations must be valid for all kinds of substitutions.

We see that the principle of invariance of equations in respect to groups of transformations accepted as the mathematical basis of a natural-scientific theory specifies the principles of ordering human experience, constituting objectiveness of the epistemological rather than ontological plane. More concretely, the objectiveness of the results of research activity in natural science is realised through the principles of relativity which introduce, as a general rule, the independence of laws of nature from the modes of their description—from experimental conditions, the scientist's individuality, specific features of experimental devices, etc.; i.e. these principles assert the independence of the magnitudes measured from the results of measurements.

A mode of realisation of the principles of relativity is the principle of invariance of the laws of a theory relative to groups of transformations which are, as a minimum, "a mode of verifying the formal correctness of formulating laws in the form of certain equations" (48, 199). In classical mechanics, an indication of the correctness of laws or formulations is invariance relative to the Galileo transformations; in the special theory of relativity, relative to the Lorentz transformations; in the general theory of relativity, relative to the permissible transformations of the Gaussian systems of coordinates, and so on. There are also general groups of transformations valid for all natural-scientific theories. For instance, the laws of classical and relativistic mechanics are invariant relative to the Poincaré groups or the Lorentz groups of transformations. The laws of classical and quantum electrodynamics are invariant in respect to the Lorentz transformations, the calibration transformation of electromagnetic potentials, and the coordinate shift. Equations invariant relative to definite transformation groups have their own mathematical apparatus, geometry and reference frames. Thus "Maxwell's equations are the simplest Lorentz-invariant field equations which can be postulated for an anti-symmetric tensor derived from a vector field' (136, 63).

A general description of links between the formulations of a theory satisfying the demand of invariance and definite groups of mathematical transformations is achieved in the Klein programme, which interprets groups of transformations through movements of space in itself. Thus "Euclidian motions, motions in the Lobachevsky space, affine transformations, projective transformations, elliptical motions, and conformal transformations form groups and define, accordingly, Euclidian geometry, the Lobachevsky geometry, affine, projective..., elliptical..., and conformal geometry.

"The group of Euclidian motions is included in the group of affine and conformal transformations... The group of affine transformations, in its turn, is a subgroup of the group of projective transformations, while the groups of Lobachevsky space motions and elliptical motions are subgroups of the groups of projective and conformal transformations; these links between groups of transformations explain the possibility of interpreting the geometry of one space in another space" (85, 26).

Similarly, in natural science "a group of transformations in respect to which the mathematical properties of the object remain immutable or invariant, establishes a measure of community ... of knowledge—the more comprehensive the group of transformations, the more general are the properties of the object which it reflects" (62, 142). The evolution of natural science shows a tendency towards a universalisation of transformation groups: a subsequent theory makes use of a more comprehensive group of transformations than the previous theory; the transformation group used in the previous theory becomes a subgroup of the subsequent theory. The objectiveness of this tendency engenders two propositions. Firstly, the progress of natural science is inseparable from the progress of mathematics, which has to satisfy the need of each new theory for the most comprehensive group of transformations in respect to which its laws would be invariant. Secondly, the relation of inclusion between groups of transformations of mutually interchangeable theories, as well as the fact that the laws of different theories, despite their specific and unique types of invariance, satisfy a number of general requirements of invariance relative to transformation groups, are of fundamental significance for substantiating the principle of continuity in the advance of natural science—they are special forms of mutual connections between theories (55, 198).

Apart from the formal function, that of being a mode of verification of the "expressive" correctness of the formulations of a natural scientific theory, the requirement of invariance of laws relative to groups of transformations also performs meaningful and heuristic functions in natural science, orienting natural science towards the search for new laws of motion satisfying, so to speak, preplanned requirements of invariance relative to definite groups of transformations. Let us choose as our illustration the Heisenberg nonlinear theory of elementary particles, in which "the form of the fundamental equation (law) was defined ... precisely on the basis of the requirement that it be invariant not only relative to spatial and Lorentz rotations but also with respect to the specific transformations of Pauli—Jürsey and Salam—Toushek, which are characteristic precisely of the modern theory of elementary particles" (48, 199).

To sum up. The requirement of invariance of the formulation of a natural-scientific theory relative to groups of transformations ensures the reproducibility, uniformity, identity, and repetition of the results, their independence from reference systems; it guarantees the objectiveness of natural-scientific knowledge, and performs the heuristic function of goal-setting.

Having discussed the reasons for the possibility and desirability of employing mathematics in natural science, let us point out some factors that impede or slow down the process of its mathematisation.

(a) Mathematisation (including quantification, metrisation, logification) is only possible in the presence of a feedback, of a creative dialogue between mathematics and natural science. Such a dialogue, however, is not always feasible. Indeed, the formulations and propositions of mathematics are meaningless on the natural-scientific plane—factually, experimentally meaningless. The formulations of mathematics are endowed with a natural-scientific meaning through interpretation, through operational definitions. In some cases, this operation of ascribing natural-scientific meaning to mathematical propositions succeeds: cf., e.g., the affine connectivity coefficient in gravitation theory. In other cases this goal proves impossible to achieve; thus a natural scientist cannot see the meaning of Archimedes' axiom, irrational and transcendental numbers, incommensurabilities, etc. In these cases, mathematisation of natural science is apparently impossible.

(b) Mathematisation is only possible where natural-scientific demand and mathematical supply are in balance. Natural scientists are known to have said ironically that a mathematician can do something, but of course not that which one wants him to do at the given moment. The situation described by the most offensive part of this dictum (the part beginning with "but") is not offensive at all—it has deep roots in the functioning of mathematics and natural science.

Three cases can be distinguished in the uses of mathematical apparatus in natural science.

(1) The apparatus of mathematics is actually used in natural science.

(2) The apparatus of mathematics can potentially be used in natural science. There are two possibilities here. (a) There exists a material possibility of potential application of mathematical apparatus in natural science. This case describes the familiar facts of the temporal gap between the two phases of natural-scientific activity—constructing a formalism and interpreting it. (b) There exists a formal possibility of potential application of mathematical apparatus in natural science. This case describes the fact of the existence of mathematical structures which are not actually used in natural science in view of their incompatibility with the demands of "substantive logic", but which, being self-consistent, can yet be introduced in natural-scientific theories.

(3) No mathematical apparatus of the kind required by natural science is available. The possibility of this case is determined, on the one hand, by the saturation of natural science with available mathematical structures which no longer satisfy its demands, and on the other, by the actual absence of the necessary mathematical apparatus which could satisfy the needs of natural science. Why does a gap between the production and consumption of mathematical structures by natural science appear? The answer is simple. Mathematics is not an appendage

of natural science, it is an autonomous science which does not in general produce results with natural-scientific meaning; that is not the objective of mathematics. Generation of results in mathematics is not subject to external demands, including the demands of natural science, but to the inner logic of changes in the problem domain. Mathematics can only satisfy the needs of natural science in the solution of those questions which can be realised as mathematical questions. The existence of this gap is thus a consequence of objective interrelation of two independent scientific disciplines. Natural science only applies mathematics when it can project mathematical structures onto its own problem domain, translating the formulations of mathematics into its own language, and vice versa.

At present, the demand for elaborating a new mathematical apparatus has increased in connection with the need to describe and analyse discrete processes. Such a need exists, for example, in physics, where the contradiction has to be eliminated between the continuity of the process of distribution of neat and its molecular-kinetic nature; between the continuity of field and its quantisability; similar problems exist in gas dynamics, and other areas. Mathematics, however, has not yet elaborated any methods for achieving these goals. Thus, the demand exists; its potential satisfaction, connected with obtaining effective results in the course of the immanent progress of mathematics, will be the content of future mathematisation of natural science.

(c) Mathematisation is only possible when natural scientists and mathematicians know what they want from each other.

The mathematisation of science is an instrument of its progress, as it makes cognitive situations explicit, facilitates the study of the object, provides perfect methods for orderly arrangement of results, promotes the growth of knowledge, and ensures the effectiveness of experimental research. There are some exceptions from this generally valid proposition.

We shall not consider here descriptive sciences founded on empirical principles; an appeal for their mathematisation would be meaningless. Let us consider theoretical sciences in which there is a demand for formalisation, axiomatisation, deductivisation, introduction of methods of mathematical modelling, mathematical planning of experiments, mathematical hypotheses, etc.—in short, a demand for mathematisation. What are the premises for effective implementation of such a programme in this case?

We know that much of theoretical knowledge has not been mathematised. How can this be explained? The reason lies, in our view, in the absence of precise knowledge of what must be mathematised and in what way. Thus there have been numerous appeals for the mathematisation of biology. Unquestionably, there are good grounds for these appeals; let us ask ourselves, however: why does biology need mathematisation? Why wasn't biology mathematised in the past? Apart from references to historical traditions, the time that was needed to achieve theoretical maturity, etc., it is argued that biology was not mathematised because of the absence of mathematical apparatus answering its substantive specificity.

Let us try to find out what demands of biology must be satisfied by mathematics and why they cannot be satisfied by available mathematics.

Let us turn to the original sources. Arguments concerning historicity, "progressism", integrality as attributes of the biological substratum that are not amenable to mathematical treatment have always been plentiful. A great many biologists and philosophers, beginning with Goethe, have expressed themselves on the subject. Leaving aside the views of avowed irrationalists like Henri Bergson, Wilhelm Dilthey and others, let us evaluate the views of their antipodes. Thus Vladimir Vernadsky writes: "Goethe, a naturalist, a precise observer and experimenter who rejected number and causal explanation of natural phenomena ... is undoubtedly right in one particular respect: causal, numerical links do not cover everything that is observed in exact natural science", for "the analytical device of dividing phenomena always leads to an incomplete conception, for in reality 'nature' is an organised whole" (16, 272).

In our view, these and similar views based on too strong and too vague assumptions cannot serve as a platform for a critique of mathematics and a negation of the possibility of mathematisation of biology. Let us demonstrate this point.

The problem of expressing in mathematical language the focal biological concepts of development, evolution, progress, etc., is a real problem. Development is a unidirectional, asymmetric, anisotropic process, whereas the laws of mathematics are symmetric about the sign of equality. Does this fact entail the impossibility of the mathematisation of biology? No, it does not. The problem of reflecting development (motion) in rigid concepts is a general epistemological problem. It emerges not only in biology but also in any science, everywhere where language, thinking, *ratio* are used. That is a universal problem of human cognition—the problem of interrupting the continuous, of dividing the indivisible, of schematising the unschematisable, of identifying the unidentifiable.

This is a fundamental but not an unsolvable problem. Proof of this is found in dialectical materialist epistemology, which interprets the process of identifying the unidentifiable as the basis of the reflective process: being adaptive in character, this process facilitates rather than impedes the adequacy of cognitive activity.

Rejection of the dialectical materialist interpretation of the role of this process is fraught with the danger of the blind alleys of epistem ological irrationalism propounded by neo-Kantians (of the Baden school), by the followers of the philosophy of life, existentialism, and other thinkers whose general position is summed up in Jaspers' proposition that identifying cognition leads to error if it is directed at a developing whole (149). In the light of cognitive achievements, the untenability of this position is obvious. It would be impossible to reject the principle of identifying the unidentifiable; that would mean a rejection of science itself, and that is, in the final analysis, precisely what the irrationalists advocate. But what is in this case the meaning of the critique of mathematics which applies, in the most explicit form, this principle? How can and must mathematics be modified in order to reflect development, etc.? Inasmuch as the question of the mathematisation of biology is considered in connection with this problem, it is obviously a serious one.

Careful evaluation of this problem reveals its methodological status, so that its solution must also be methodological. We know from methodology that mathematics is a very strong apparatus of identification, its foundation resting on the relation of identity central to mathematics. The reliance on this relation makes mathematics tautological, to an extent, as Frege, Poincaré, Meyersohn and others pointed out. They stressed, however, that the tautological nature of mathematics is overcome because in the course of identification the nonidentity of that which is identified is clearly realised.

Thus recourse to regulators of epistemological rank makes the situation less acute, imposing the view of the result of identifying the unidentifiable as requiring further concretisation and specification rather than as final and frozen. The solution that lies within the boundaries of the dialectical approach to the process of cognition, to the evaluation of its results. In this case, the question of the inadequacy of mathematics no longer arises.

The mathematisation of natural science is often impeded by the absence of a clear reflective position as to what particular elements of natural science must be translated into the language of mathematics and how this translation can be achieved. In particular, the serious problems arising in connection with the mathematisation of biology can largely be eliminated by clarifying the semantic basis of the arguments, by transposing the discussion of the topics involved from a vague, general plane onto a definite and concrete one. Let us draw a parallel. On the one hand, mathematical set theory is not free from inconsistencies. On the other hand, the method of studying contradictions is, by its very essence, dialectics. The conclusion thus suggests itself that there is a need for introducing the dialectical element in set theory (14, 166).

The question arises, however: How is this element to be introduced? What must be done to achieve this objective? There are no suggestions for this—for the simple reason that, in our view, no such suggestions can be given on this approach. Methods for eliminating antinomies in set theory should be sought for in mathematics itself—either on the traditional basis of the familiar programmes for the foundations of mathematic or on the non-traditional basis of constructing paraconsistent logics. In any case, appeals to dialectics are inappropriate here. Dialectics does not provide concrete instructions for the solution of intrascientific problems; it acts as a heuristic—methodological, normative, worldview—basis for special scientific research.

To go back to biology: the mathematisation of this science is hindered by vague talk about the allegedly irrational, mathematically inexpressible nature of its reality. The fact that biology deals with a special kind of reality is obvious. Thus the behaviour of non-equilibrium, non-stationary, self-organising open systems—the kind of systems to which biological objects belong—differs from the behaviour of mechanical systems. However, the problem does not lie in the fact that biological systems are characterised by integrality, development, etc. The problem lies in the existence of extremely vague ideas of integrality, development, etc., in general and in the absence of special conceptions of integrality, development, etc., of a concrete (biological) level. In this connection, wide prospects are opened up, we believe, by synergetics, which provides a properly scientific model of development. The heuristics of the synergetic approach comprises such natural-scientific concepts as mass, energy, structure, etc., which permit the modelling of certain aspects of biological development in terms of increased complexity of structure, growing levels of organisation, and other features resulting from mass-energy exchange processes. The question of mathematical servicing becomes much clearer here. The tasks of mathematics are in this case specified as the elaboration of an apparatus for expressing development as a discrete process of successive replacement of qualitative states resulting from leaps from one level of organisation to another. Mathematics possesses such an apparatus—namely, the theory of catastrophes. Mathematisation thus proves feasible. Bearing in mind the possibilities of the systems approach and of the mathematical apparatus conceptually connected with it, the same can be said of the concept of integrality.

Thus mathematisation is only possible when natural scientists and mathematicians know precisely what they want of one another.

In many instances mathematics knows exactly what it can and wants to give to natural science. Frequently, however, there is no such knowledge—mostly because mathematics is reactive to the substantive needs of theories mathematised. Must a mathematician be also a specialist in natural science, to achieve its mathematisation? As often as not, there is no such need. For example, at all levels of the organisation of matter, regardless of the morphological or structural specificity of the objects, their kinetics is universal; that is why it can be modelled without much hindrance; the mathematician requires no special knowledge here. In other cases, however, a mathematician engaged in mathematising a natural scientific theory does need special training. It is pointed out, for instance, that one of the causes of lack of progress in the mathematisation of chemistry is the fact that only five per cent of mathematicians are conversant with the problems in this field. It follows that, for the modelling of natural-scientific reality to be successful, mathematicians must study this reality on the meaningful level.

To achieve their effective mathematisation, the substantive features of natural-scientific theories are taken account of in the framework of the so-called substantive axioms. For instance, a general approach, devoid of concreteness, to the mathematisation of biology does not permit the choice between discrete and continual systems of thinking. This choice, however, is quite feasible on the concrete approach. For example, the continual style of thinking is preferable in the mathematisation of morphology, while the mathematisation of genetics requires the discrete style. All this must find support in substantive axioms, which can act as a kind of mediators between substantively neutral mathematics and substantively oriented natural-scientific theories. The uniqueness of biological systems may be effectively interpreted by the theory of categories, capable of operating with the local time of the systems involved, of functor comparisons of structures, etc., provided individuality, mutability, etc., are specified by suitable substantive axioms. There is a potential here for building bridges between biology and physics, bearing in mind that progress in the latter, according to John Wheeler, will consist in the transition from relativity to mutability (192, 242).

It may thus very well be that new axioms rather than a new mathematics are needed for the mathematisation of many as yet non-mathematised theories.

The adoption of empirical grounding as the central criterion of scientificity, which covers fundamental observability and experimental justifiability.

The requirement of fundamental observability was implicitly introduced by Einstein in his critical evaluation of the nature of such basic physical concepts and principles as absolute rest, remote action, etc., which were shown to be pure fictions. We owe explicit introduction of this principle to Werner Heisenberg, who formulated the rule of exclusion from quantum mechanics of the classical concepts of orbit, trajectory, etc., which have no empirical analogues there. In the course of time this requirement assumed the status of the methodological principle that concepts which refer to distinctions beyond possible experience have no physical meaning and ought to be eliminated (120).

The question now is, what is the real content of this principle? If we were to verify theories on the basis of this principle, "we would have to reject all molecular, atomic and electronic theories, ... the theory of relativity, the quantum theory; ...in fact, almost the entire new physics would be gone, as well as a great part of the old physics" (165, 168).

Does the principle of observability implement an untenable empiricist ideal? Let us stress that the principle of observability is polysemantic. It can be given the following interpretations: (1) Direct observability, implying the finding of operational criteria and procedures for the identification (definition) of the empirical meaning of propositions. (2) Observability in principle, the content of which can be reduced to the search for the laws of exclusion rejecting the possibility of observing an object (e.g., of any contradictory object). A defect of this demand is, perhaps, the fact that we do not know all the laws of exclusion that can restrict scientific inquiry. (3) Experimental verifiability, reducible to empirical substantiation of hypothetico-deductive theories.

The narrow empiricist, physicalist ideal of scientificity raises the first interpretation of observability to an absolute. On the positive side, that principle requires maximal substantiation of the ingredients of a natural-scientific theory by experimental data. Any other interpretation—as, e.g., the literal interpretation, "the elimination of all non-observables from theory"—leads to nonsense (119, 18).

Clarification of the nature of the principle of observability as a methodological imperative must be guided by two basic assumptions: (1) "the decision as to whether a given physical magnitude is in principle observable ... can never be arrived at a priori but always from the standpoint of a definite theory only" (178, 201); (2) as opposed to the operationalist viewpoint, there is no universal criterion, the same for all theories, that might determine whether a proposition has, or has not, an empirical meaning.

Not all ingredients of a theory must answer the demand of observability. Besides, what is observable in one theory may be unobservable in another. In quantum mechanics, for instance, there are such unobservable entities as wave function, Hilbertian phase spaces, etc., which have no explicit operational meaning. In its most precise sense, the principle of observability "refers to concepts which had operational meaning (even if it were only an indirect meaning) in an old conceptual system but lose it in the new one" (5, 101). Of this nature are many classical concepts which were given a more precise formulation in the course of time. Thus in experimenting with atomic phenomena, quantum physicists continue to use classical concepts, but their application is limited by the uncertainty relation.

The demand of experimental justifiability, which means potential experimental verifiability (or falsifiability) of systems of natural-scientific knowledge, acts as an empirical criterion of their scientificity. Criteria of scientificity other than the empirical are also used in natural sciences, in particular logical and non-empirical, which are not, however, the primary ones.

In natural science, whose objective is adequate reflection of reality, the most fundamental criteria of scientificity are the empirical criteria, which directly verify the perfection of natural science precisely from this point of view. The fact that the empirical criteria of scientificity are of fundamental significance for substantiating natural science points to the existence of a close link between the definiteness of the correlation between the B_c of a theory, with its corresponding B_e, and the concept of scientificity. In regard of these correlations, a theory is believed to be unscientific or epistemologically undesirable if (a) it is confirmed by the entire universe of the factually given (the omniscience syndrome); that is only attainable if the theory in question contains absolute truth (which is impossible, for obvious reasons), or if it is potentially irrefutable (in this case we have an arbitrary fantastic natural-philosophical construct); (b) it has no links with empirical facts and cannot in principle be confirmed; in this case we have an artificial assumption (e.g., the Lorentz—FitzGerald assumption); (c) it explains but does not predict, being a description *a tergo*, a semantically trivial construct (as e.g. Driesch's concept of entelechy); (d) it predicts but does not explain existing laws (we then deal with a surmise or conjecture); (e) it both explains and predicts, but the predictions are not borne out by the reality.

The proposition that, if a theory (T) has observable consequences (O), the verification of T is reducible to the search for O, requires some explanation. The point is that substantiation of the truth of T is not a narrow empirical procedure for finding O only. Confirmation of T taking into account the rules of derivation in classical logic does not yet permit to derive the truth of T from the truth of $(T \rightarrow O) \cdot O$. At the same time confirmation of T by the actually discovered O is an element of the substantiation of its truth. As for the refutation of T, which follows deductively, according to the rule of *modus tollens,* from the truth of the conjunction $(T \rightarrow O) \cdot \sim O$, it is also insufficient for final conclusion about its falsity, although it is an element in the substantiation of its falsity. The assumption that the *modus tollens* is radical is too strong, as it does not take into account many essential realities of science imposing limitations on automatic use of this logical rule in scientific practice.

To clarify our idea, let us stress the fact that the interconnection between B_c and B_e is far from single-valued, and that the latter term is only ultimately the decisive and final instance in the substantiation and verification of the former one. The realisation that there are discrepancies between B_c and B_e does not result in immediate discarding of B_c; it assumes close study of the composition of B_c to elucidate the following questions: (1) How does B_c describe, explain and predict phenomena in B_e? (2) What resources in B_c are used to achieve this? (3) How does B_c agree with the existing system of knowledge, with the picture of the world, and the general cultural background?

What types of contradictions are permissible in natural-scientific theories, and which of them ultimately lead to a crisis in these theories? Only logical contradictions are impermissible. Contradictions between the facts of a theory *are* permissible, but these are smoothly eliminated in some cases while in others they lead to a critical situation which entails a scientific revolution. In what sense and why are factual contradictions in theory permissible?

A theory is an integral system of mutually connected propositions which, taken in their unity, are correlated with reality, describing and explaining it. These theoretical propositions taken in their totality are the instruments of carrying out and explaining experiments, of identifying and forming scientific facts, and of conceptualising the empirical basis assimilated by the theory. When contradictions between the facts of a theory are revealed, it is impossible to say which particular theoretical proposition leads to this contradiction; we must analyse the interrelation between theory and the empirical facts as a whole. The set of theoretical assumptions is in such cases often extended to include further assumptions and modifications facilitating the elimination of the contradictions of the facts of the theory. This latter move is justified by the Duhem—Quine principle; from the integrality and systematic cohesion of theoretical scientific reality, Quine concluded that "any statement can be held true come what may, if we make drastic enough adjustments elsewhere in the system" (179, 43).

Do we know which component of theory leads to contradiction between theory and empirical facts? The greater the number of auxiliary assumptions in the theory, the more difficult it is to answer this question. If the structure of a proposition to be verified is presented in the form

$$(H \cdot J_1 \ \ldots \ \cdot J_n),$$

where H is a hypothesis; and J, a set of assumptions, the formula

$$[(H \cdot J_1 \ \ldots \ \cdot J_n) \rightarrow O] \ \cdot \sim O \rightarrow (\sim H \vee \sim J_1 \vee \ \ldots \ \vee \sim J_n)$$

shows that both H and any of the J's can lead to a contradiction between theory and experiment, and it is impossible to say which of the J's has this effect.

The admitted possibility of contradictions between the facts of a theory rests on the following. (1) Facts cannot correspond to theory with absolute precision: theory operates with idealisations, conceptual and logico-mathematical structures, whereas being reflected in theory is non-ideal and non-mathematical. (2) Theory always reserves the right to conceptualise the fact contradicting it, which may eliminate the contradictions.

(3) Allowance must be made for errors in computation, measurement and calculation at the empirical level.

The first proposition is self-obvious. The content of the second proposition may be explained by the following example. One and the same fact—the absence of deviation to the west of a body falling from some height—is differently conceptualised within the geocentric and heliocentric frameworks. Thus adherents of geocentrism regard this fact as proving the immobility of the Earth, since its assumed motion would produce, in their view, accelerations unconnected with the interaction of bodies, which would determine the deviation effect. To refute Ptolemy's argument, Copernicus introduced the concept of "natural motion" which does not interfere with the natural order of processes on the Earth; interpreting the absence of deviation to the west of bodies falling from a height from this standpoint, he stated that this fact does not contradict heliocentrism.

Analysis of the third proposition is probably of the greatest interest in the discussion of the permissible limits of error in experiments intended to prove or disprove a theory. In some cases, the discrepancy between theoretical conclusions and experimental data is regarded as an entirely ordinary and negligible event, whereas in others it is seen as an extraordinarily important event demonstrating the untenability of the theory in question.

Already in the 17th century, Newton suggested that the Earth is a spheroid, but he failed to prove his hypothesis. It was only done in the 18th century by Alexis Clairaut, who constructed the theory of the Earth's non-uniformity. When empirical calculations were made, it was found that they did not accord with theoretical conclusions. This is what Clairaut wrote of this fact: "As the measurements ... were executed with great exactness and considerable attention, it appears at first sight that these measurements must be preferred to those of my theory; however, if one pays attention to the errors inevitable in the actual measurements, and to the limits of these errors, one will see that, without blaming these measurements, one can bring them closer to my theory, and find a common result" (129, 299). Thus Clairaut assumed that he could ignore the discrepancies that came to light, attributing them to the errors (insignificant, in his view) made during measurements.

And here is an example of apparently acceptable limits of such errors being exceeded, which caused a revision of the theory in question. Christopher Hansteen calculated the magnitude of the magnetic field assuming the existence of two infinitely small magnets differing in position and strength. Gauss compared the data deduced from that theory with experimental ones; the comparison showed significant discrepancies between theory and observation, great enough for Gauss to discard Hansteen's hypothesis.

Thus we see that in some cases it is believed possible to retain a theory by a kind of remission of the fact that empirical data contradict it; in others, this fact is regarded as sufficient for falsifying the theory.

What is behind all this? Why is a discrepancy between fact and theory seen as an error at the empirical level in one case, and in another, as an indication of the imperfection of theory? Essentially, it all comes to this problem: what is the measure, degree and interval of permissible experimental error, and when is it impermissible to exceed it?

It is important to stress that there is no general solution of this problem. It can only be eliminated through contextual analysis of actual situations. The point is that the interval of permissible error cannot be determined a priori. The context in which a certain error is regarded as negligible must always be specified.

For example, the hypothesis is worked out in the framework of a project for a unified field theory that there is a certain interaction uniting the electromagnetic and weak interactions. One of the consequences of the hypothesis asserts that, if the electro-weak unified interaction is real, there must be a difference in the force of interaction between protons and electrons with different spin orientations. This hypothesis was confirmed in experiments at the Stanford accelerator; indeed, deviation in electrons with counter-clockwise polarisation is $^1/_{10^4}$ greater than in electrons with clockwise polarisation. That means that electro-weak interaction actually exists.

Leaving aside problems connected with the evaluation of the polynomial tendency of the experiment, let us ask this question: what does confirmation based on a $^1/_{10,000}$ effect mean? In the times of classical mechanics, such an effect would have been disregarded and seen as an error only. It follows that an interval must be specified of the sensitivity of theory to the smallness of a concrete effect. In this case, it was this sensitivity which permitted the interpretation of the effect as confirmation, while under different conditions this effect would have been regarded as error. Each natural-scientific theory thus has its own interval of smallness. What happens if its limits are exceeded?

The theory is then subjected to an allround interdisciplinary examination—metatheoretical (analysis of inner consistency), intertheoretical (establishment of coherent perfection), philosophical (the study of epistemological assumptions), and empirical (investigation of the degree of confirmation of the predictions). Only after such an allround examination is the verdict pronounced.

It is therefore incorrect to state that a theory is rejected only because there are empirical data contradicting it. The correct position is as follows: the existence of empirical data contradicting a theory is merely a signal for an allround rational examination of the theory, which may result in retaining it without any changes, in its partial revision, or rejection. One of these cases was illustrated by the Clairaut episode, another, by the Gauss example.

Partial revision of a theory, of which the nucleus is retained, is illustrated by Johannes Kepler's reform in the framework of the classical model of heliocentrism. Originally a confirmed adherent of the Copernican theory of circular motion of the planets, Kepler once "noticed that one planet... (Mars.—*V.I.*) was eight minutes of arc off, and he decided this was too big for Tycho Brahe to have made an error, and that this was not the right answer" (138, 16). The awareness that Tycho Brahe—an astronomer who obtained the most exact results, as a rule—could not have made such a considerable mistake compelled Kepler to modify the elements of circular trajectories used in the theory, introducing elliptical orbits, which eliminated at once the discrepancy between theory and facts.

To complete our analysis of the problem of discrepancy between B_c and B_e, let us stress that radical verification or falsification of theory by experience is impossible. This stems from the nature of activity at the empirical level, which tends, in the framework of the procedures of confirmation and refutation, towards polynomialism. A general evaluation of the nature of this activity yields the following conclusions.

Experiment can confirm different theories complementing one another (cf. the simultaneous experimental confirmation of the corpuscular and wave theories of light).

Experiment cannot form a basis for preferences as far as the truth of one theory as against another, competing theory is concerned (cf. the competition between the ether theory with the Lorentz—FitzGerald modifications to compensate for its defects, and the special theory of relativity).

Experiment can confirm a false theory, which follows from the specific properties of logical derivation; according to the rules of implication, falsity cannot follow from truth but truth can follow from falsity. Thus Newton's false theory of sound agreed with reality as far as the magnitude of sound velocity was concerned, and that made it difficult to criticise that theory.

Results of experimental activity are rationalised and conceptualised, which makes an *experimentum crucis* impossible. Theory cannot be reduced to observation propositions; empirically verifiable propositions derived from theory do not embody the whole of its content; they are not absolutely strict and reliable. There are no pure observation records, no immutable and hard empirical basis of knowledge which classical empiricists dreamed of and believed possible.

Experiment is not guaranteed against error, the more so that verification of many theories, such as the general theory of relativity, requires extremely precise measuring techniques.

There is a difficulty, called the confirmation paradox, which complicates the task of optimising the procedure of confirmation. The confirmation paradox arises if we assume that (a) the propositions (hypotheses, laws) of a theory are confirmed by the entire universe of data that do not contradict them; (b) the data may confirm any logically equivalent propositions (hypotheses, laws). The paradox consists in that the facts confirming a certain proposition include, among others, facts which the proposition in question does not mention explicitly. The paradox arises in view of the following considerations. If H is a hypothesis of the type

$$\forall (x) \; (P(x) \rightarrow Q(x)) \qquad (1)$$

then, since (1) is logically equivalent to

$$\forall (x) \; (\sim Q(x) \rightarrow \sim P(x)) \qquad (2),$$

H is confirmed both by (1) and by (2). Thus the universal proposition "All men are mortal" is confirmed by the proposition "x is a mortal man" and the proposition "y is an immortal non-man".

Karl Hempel called the paradox of confirmability of universal propositions by any facts unconnected with them a pseudo-paradox, arguing that, since universal propositions of the type "any P is Q" exclude all objects having the property P but devoid of the property Q (143), the set of objects regarded as potential verifiers of universal propositions must be restricted. Elimination from this set of objects irrelevant for links between P and Q results in the impossibility, properly speaking, of "confirming universal propositions by statements about objects not mentioned in the propositions themselves" (112, 77).

The confirmation paradox arises in connection with the logicist approach to the procedure of confirmation. It is usually interpreted as "an argument which can, if only partially, fill the gap between the universality of the verified hypothesis and the limitations of the facts confirming it. However, a purely logical expansion of the set of facts involved in the procedure of confirmation contradicts the actual scientific process" (ibid.). There is a real possibility of overcoming the blind alleys of the logicist approach to the study of confirmation; it lies in the practical approach worked out in dialectical materialism, to the analysis of this procedure. However, the practical approach—a fundamental method of solving the problem—proves to be too general to be an effective instrument of confirmation of concrete individual hypotheses as they occur; however, this does not cancel the task of optimising the procedure of confirmation of universal propositions in terms of a limited ensemble of empirical data.

Experiment is not an instrument of meaningful evaluation of the heuristic potential of theories; this follows from the fundamental experimental verifiability of both the original and bold theories stimulating progress in science and the ineffectual *ad hoc* constructs slowing down scientific development.

There exists a general epistemological problem of justifying induction, of substantiating the necessary character of inductive conclusions having a non-demonstrable, factual status. The familiar programmes for substantiating induction—deductive (John Stuart Mill, John Keynes), pragmatist (Charles Pierce, Hans Reichenbach), inductive analytical (Rudolf Carnap, Kaarlo Hintikka), conventionalist (Henri Poincaré, Edouard Le Roy), and linguistic (Alfred Ayer)—are none of them fully adequate. Rationalists, and critical rationalists in particular, therefore felt justified in describing experience as an inadequate instrument of attaining the truth, and knowledge obtained with the help of this instrument as unobligatory, unreliable, "doxophic" and thus unscientific; true science was limited to demonstrative logico-mathematical knowledge.

The rationalistic critique of experience and induction as "untenable" instruments of cognition is itself untenable. First, it relies on an inadequate conception of absolute metaphysical truths, which cannot (unless of course we postulate their supernatural origin) be arrived at either through experience or by applying any other cognitive instruments available to man. Second, it permits the unacceptable antidialectical interpretation of deductive knowledge as absolutely certain. As we have pointed out above, the Achilles' heel of deduction as a method of generation of knowledge is "non-obvious" axioms and other "opaque" propositions, rules and laws accepted without proof—the axioms of choice, the theorems of pure existence, the rule of excluded middle, etc.

Experimental, inductive knowledge is relative, but its relativity is not tantamount to its non-objectiveness. Relativity is an indication of the historical conditioning, a characteristic of the limits of approximation of the truth by knowledge. The rationalist assertion of the epistemological untenability of induction, of experimental knowledge, is therefore basically wrong. Strictly speaking, induction is just as reliable a method as any other, and knowledge obtained by induction is just as effective as knowledge obtained by any other method used in science. This view is substantiated in the programme of practical and mediated justification of induction developed in dialectical materialism. There are certain problems here as well. Practice only guarantees the necessity of inductive conclusions as a tendency, which determines the probabilistic status of concrete inductive conclusions, distinguishing them from clearly apodictic, demonstratively obtained conclusions.

In view of this, the polynomial tendency inherent in experience does not permit any single-valued substantiation of the truth of theories in terms of their empirically confirmable consequences. The same can be said about the procedure of substantiation of the falsity of theories in terms of their empirically rejected consequences.

As we have pointed out above, the illusion of the existence of an *experimentum crucis* providing "absolute" counter-examples of a theory is at present dispelled. Any practising scientist, with his obsessive faith in the correctness of the theory he works on, interprets such counter-examples either as fluctuations or as stimuli for a partial revision of the theory intended to improve it. Generally speaking, falsification cannot be an instrument of unequivocal and radical rejection of theories for the following reasons.

In view of the systemic nature of knowledge, refutation of derived hypotheses is merely evidence of the falsity of certain elements of a theory, and not of its falsity as a whole.

Contradictions between facts and theories can be eliminated by compensatory *ad hoc* modifications.

Theories can fully retain their fundamental laws even if the untenability of their interpretative and explanatory apparatus becomes apparent—by being included in new and more adequate theories. For confirmation let us refer to Descartes' theory of light, which categorically insisted on the instantaneous character of the propagation of light. When Olaus Roemer calculated that the velocity of light propagation is finite, the general impression was that Descartes' theory collapsed at a stroke. However, the downfall of this theory did not affect the laws of light refraction—they remained intact as part of elementary optics.

A theory is rejected when it is established that (a) no modification of the theory brings harmony into its relationship with empirical facts; (b) modifications of the theory are artificial in character, entailing unjustifiable complexity in the theory; (c) there is a new theory successfully competing with the old one.

The procedure of falsification is thus not as radical as critical rationalists would have it.

Experience does not unequivocally guarantee the truth of a theory—for the following reasons.

Identical B_e are compatible with different B_c. Extremely characteristic in this sense is the situation in cosmology, with its proliferation of theories alternative to Fridman's theory of an expanding universe, although there is no pressing need for these alternative theories, as Fridman's theory adequately describes empirical data, makes experimentally confirmed predictions, etc. The reason for the proliferation of theories in this particular case is not clear, but the nature of the many-valued correlation between B_e and B_c, manifested in the one-to-many principle of their relationship, is obvious—it lies in the hypothetico-deductive schema itself of the unfolding of natural-scientific knowledge at the theoretical stage of its existence. We can take it as proved that modern natural science is built on an alternative-complementary (pluralistic) basis. Illustrations can be drawn from biology (tychogenesis vs nomogenesis), geology (fixism vs mobilism), or the general theory of relativity (the metrical theory vs the theory of direct interparticle interaction).

In many cases experimental verification of natural-scientific knowledge is difficult or impossible, as in palaeontology, petrology, soil science, climatology, seismology, astronomy, cosmology, etc. For fundamental or special reasons, many propositions of natural science cannot be experimentally substantiated. For example, the hypothesis that the gene is a protein molecule, had no experimental foundation at one time and was only confirmed theoretically, being in agreement with the general cultural background. The Avogadro hypothesis, accepted in science, is experimentally non-verifiable; it was accepted for reasons of "inner agreement that is established between various experimental data" (108, 58), such as the laws of gases or the Dalton theory. All these factors make it necessary to use non-empirical criteria of scientificity in natural science. If it is a question of an "absolutely new" theory, it must satisfy the demand of coherence, that is, of compatibility with a practically verified system of accumulated human knowledge. If it is a question of mutually interchangeable theories, with partially intersecting classes of referents, the new theory must satisfy the requirement of correspondence.

And yet, as the following argument shows only experience can be the criterion of the truth of a natural-scientific theory.

The categorial apparatus of a natural-scientific theory "is not created for the system; the system together with ... the apparatus is constructed for the description of a certain domain... The choice of a conceptual system is not arbitrary in the sense that theories must give essentially one and the same answer about identical things (the question requiring different conceptual apparatuses in different theories)" (69, 176). Max Planck was quite right therefore when he called facts "the Archimedean point from which even the most weighty theories can be shifted" (178, 67). Facts do not form a theory; to be more precise, they are insufficient for this purpose; but theories are constructed for the facts in relation to which they perform their cognitive functions. Facts are more fundamental than theories. "In every science," wrote Engels, "incorrect notions are, in the last resort ... incorrect notions of correct facts. The latter remain even when the former are shown to be false. Although we have discarded the old contact theory, the established facts remain, of which this theory was supposed to be the explanation" (59a, 160).

Experiential (sensuous, experimental) cognition forms a necessary component of any research act of natural-scientific activity. There is no way of linking up the formulations of science with the reality which it reflects other than the path of experiment, of active manipulation of the object with the aid of empirical-level instrument—from visual observation to structural decomposition of the object.

In view of all this, we regard empirical substantiation the central, if not the only, criterion of the scientificity of natural science.

3.3. TECHNICAL KNOWLEDGE

The subject-matter of technical knowledge. One of the most important characteristics of knowledge is reference to the domains and objects with which its concrete forms are correlated. The problem of subject-matter is a topical one for all kinds and forms of cognition.

This problem has a special significance for technical knowledge. There are several reasons for the emphasis on the subject-matter of scientific-technical knowledge in epistemological studies: comparatively recent emergence of the technical sciences, intense development of technical knowledge at a time of the scientific and technological revolution, and the unending process of the formation of new technical sciences. These and other factors (such as the successive replacement of the objects of research in the technical sciences) continually revive the question of the subject-matter of scientific-technical knowledge in the methodology of scientific cognition, as well as the question of its correlation with the subjects of other forms of scientific cognition.

Traditionally, the question of the subject-matter of scientific cognition is included in epistemology, in the methodology of science. That does not mean that specialists in some other branch of science cannot effectively discuss the problem of the subject-matter of their field. On the contrary, major scientists specialising in concrete fields of knowledge necessarily feel the need for epistemological, methodological reflection on the subject-matter, methods and tasks of the field in which they work. As a rule, this happens at times of revolutionary transformation of this field or at any rate at a time of a qualitative leap in its development.

Methodological reflection on the subject-matter of scientific knowledge takes place against the broader background of methodological contemplation of the nature of scientific knowledge, its tasks, methods, etc. Concern for methodology is quite natural for modern cognition, not only scientific but also, say, artistic. Moreover, methodological studies in scientific knowledge increasingly become a necessary condition of producing new ideas in science itself (20, 59; 64, 147). This observation is valid not only for natural science, mathematics, etc., but also for human knowledge in its scientific and "extrascientific" (*inonauchnaya*) forms, which we shall discuss later in the section on the human sciences.

At present, methodological analysis of the subject-matter

of technical knowledge is closely linked with methodological research in non-classical (interdisciplinary) scientific-technical fields which are, in their turn, closely connected with engineering and design. Correct methodological interpretation of the problems of non-classical technical disciplines and non-classical engineering (as well as its classical form) is a necessary condition of developing a correct conception of technical knowledge—its organisation, functioning, and subject-matter. Only an adequate methodology of both classical and non-classical forms of scientific-technical knowledge and engineering opens up the perspectives for the development of a correct view of technics as the subject-matter of technical knowledge.

Insufficient attention to the methodology of engineering and design and lack of studies in the specificity of engineering even in its classical forms make specialists in epistemological questions of technical sciences include engineering in the domain of technical knowledge. In this case, transformation of natural materials into technical instruments, construction of technical objects, etc., are believed to be the principle task and the most important function of technical knowledge (or sciences).

This conception of technical sciences and technical knowledge in general has been criticised in the literature. Indeed, science, the technical sciences included, is an activity, but it is a special kind of activity—it is the production of knowledge taking place in the sphere of the ideal. The activity of implementing knowledge in reality, of transforming reality, takes place on the basis of knowledge but in a different sphere—in the sphere of engineering, of technological practice. However, this is only one aspect of the question. Technical sciences of the classical type must not, indeed, comprise engineering activities. But non-classical scientific technical disciplines (ergonomics, industrial design, systems engineering, etc.) embrace not only designing but also the introduction of technological systems in practice.

In view of the close links existing at the present stage between engineering and the scientific and technical disciplines, the problem of their methodological analysis presents two aspects of one and the same issue. The fact that engineering and design have been little studied in the Soviet literature is mostly due, in our view, to analysis of engineering in terms of the technical sciences. The technical sciences emerge as a result of merging of natural science and of technical practice, that is, of engineering practice at the initial stage of its development. Engineering practice as the activity of applying scientific knowledge to production, to technology, has not been identified as a special object or studied in the philosophy of technology. Engineering activity, the introduction of scientific knowledge into production, the objectification of knowledge has been considered in the context of the functioning of technical sciences. At present, there is an urgent need for a methodological analysis of engineering activity as such, of its genesis and specificity, its differences from other forms of practice. The growing interest for this problem in Soviet literature has also borne fruit in the clarification of the subject-matter and genesis of the technical sciences. We shall return to this question below.

The question of the subject-matter of technical knowledge is always linked in the methodology of scientific cognition, in epistemology, with the study of the relation between technical knowledge and technics. In analysing this relation, the idea is often expressed, in one form or another, that technics is the subject-matter of technical knowledge. Such a standpoint is natural, but methodologically it is merely the first step towards clarifying the subject-matter of technical knowledge. Consistent development of this starting point involves complex and largely unsolved problems.

It is probably a lack of clarity about the prospects and the paths of the solution of these problems, that engenders attempts to give a different theoretical interpretation of the problem of the subject-matter of technical knowledge. Of the greatest interest in this respect is the approach based on analysis of activity—of man's practical activity involving objects. The latter is considered on two planes—subjective and objective. The subjective element is constituted by the subject and his actions with objects. The objective aspect is represented by objects included in the acts of activity and interacting with each other.

This activity-oriented approach to the study of technical knowledge is connected with the use of such debatable polysemantic concepts as practice and activity, object and subject-matter of cognition. The pairs of terms connected here by "and" are often used as synonyms. Despite a certain community of these categories, they also have real differences in meaning and the classes of objects they comprise. In different contexts, these differences may lead to a lack of agreement.

In particular, the emphasis in this approach is on practical activity involving objects, which naturally comprises a narrower class of phenomena and processes than activity in general, not to mention practice. From this standpoint, the object (or subject-matter) of technical knowledge is the objective structures of this activity. But this definition of the subject-matter (or object) of technical knowledge inevitably entails another, to the effect that the object of technical knowledge in the broad sense is man's objective practice interpreted as a unity of the subjective and objective aspects. Thus it is no longer the objective structures of practical activity or even of practice as a whole that are the subject-matter of technical knowledge but the whole of practice in the unity of the subjective and objective aspects.

Neither this way of reasoning nor the conclusion are accidental. Their premises are contained in the interpretation of the objective and the subjective constituents of activity, in the conception of the subject-matter and object of knowledge. In this conception, the objective constituent of activity is characterised by the totality of objects which are connected by definite mutual relations in the separate instances of activity. In view of the object-related and goal-directed character of human activity, the relations of objects in the objective structure are obviously determined by a certain goal. The subject's goal-directed actions performed upon objects in the process of activity become "extinguished" in the objects, placing them in definite mutual relations, making them goal-directed and subjective. The objective structure of activity is thus transformed into a unity of subject and object. Continued existence of the objective structure of activity in a pure, subjectless form is impossible in view of the object-related and goal-directed character of activity. This is only possible if the subject does not deal with objects in his objective activity. Only then will the subject with his actions, their results, etc., be on one side, as it were, while the object of activity in its natural subjectless intactness, on the other.

The study of the objective structures of activity on this approach shows that their structures are determined by the goal. In other words, the objective structures of activity are goal-directed or, putting it differently, they are goal-directed structures of activity. This proposition, in conjunction with their being the subject-matter of technical knowledge, warrants some interesting conclusions. Below we shall substantiate the view that technics may be interpreted as artificial, i.e., goal-directed instruments of activity. In principle, goal-directed structures of activity and goal-directed instruments of activity are obviously identical.

On the basis of the above, we may conclude that technics can be interpreted as the subject-matter of technical knowledge in the above-mentioned sense in the context of the analysed conception as well. In other words, technics constitutes the real phenomenon on which technical knowledge closes.

An important element of an adequate and consistent interpretation of the subject-matter of technical knowledge (apart from the methodological work mentioned above), engineering activity and non-classical scientific-technical disciplines is a clear understanding of the subject-matter and object of scientific cognition in general and their concretisation in relation to technical knowledge. Technics as the subject-matter of technical knowledge (technical sciences) is, of course, a unity of the subjective and the objective (115, 6). In this respect, however, it does not differ from the subject-matter of scientific knowledge in general.

This general conception of the subject-matter of scientific cognition must be concretised in the same way as, say, physicists elaborate the concept of physical reality. To distinguish the subject-matter of technical knowledge from other forms of scientific cognition, in particular from exact natural science, we must concretise and detail the nature of the subjective and the objective in the subject-matter of technical knowledge, and to clarify their mutual connections and mechanisms of their relation to technical knowledge, otherwise a conflict arises with the principles of the theory of reflection.

As we know, Marx described machines, locomotives, railways etc. as "*organs of the human mind which are created by the human hand,* the objectified power of knowledge" (67, 29, 92). Knowledge objectified in machines apparently cannot reflect the machines before it is objectified in them. The problem arises of studying technology, its complex structure, and the correlation of that structure with technical knowledge—which, in one way or another, closes on technics.

The problem of defining the concept of technics. The word "technics" covers a wide conceptual field in the system of science and art, and generally in the domain of culture. One of the causes for that was, apparently, the original syncretic use of the term *technē* in antique culture, covering simultaneously art, science, and the crafts. The field of reference of *technē* was so wide in antiquity that it may be translated, in the context of classical culture as "goal-directed activity" or "meaningful activity" (52, 355).

The differentiation of the sciences and the arts and generally of different areas of material and nonmaterial activity in the process of historical development entailed the spreading of the term "technics" through many spheres of human culture. Its main meanings at present are the concept of technics as activity (skill, art, etc.) and technics in the narrower sense—as a designation of specifically organised material objects or systems (machines, instruments, etc.). Let us conventionally call technics in the first case "technics—activity", and in the second, "technics—object".

The starting point of the materialist interpretation of technics is the view of technics as a system of material objects. It is obviously impossible to explain the emergence and development of technics or understand its role in social life without close study of machines, instruments and other implements of labour. Analysis of Marx's theory of labour and of the elements of labour, in conjunction with the preliminary interpretation of technics as a system of implements and machines, warrants the conclusion that technics comprises all means of labour. The subsequent evolution of this viewpoint resulted in the idea, shared in fact by most Soviet scholars, that "technics should be seen as the totality of artificially constructed means of human activity" (94, 10).

This level in the development of the Marxist concept of technics, despite its relatively accomplished character, and comprehensive coverage of the phenomena involved, entails certain serious problems that have not been fully and clearly solved. Above all, the definitions formulated within this framework produce a feeling of theoretical discomfort. They can be simplified, minimised, so to speak, and logically clarified. For example, the work mentioned above interprets technics as the totality of artificially constructed means of human activity. In this definition, the concept "artificial" stands next to "man-made", both of them obviously intended to distinguish technics from the natural means of labour. "Artificial" is obviously synonymous with "man-made" (183, 6), and a correct definition should take this synonymity into account.

The category of activity, which is one of the fundamental categories in the interpretation of technics considered here, involves more difficult problems. The difficulties and differences of opinion in philosophical, methodological and psychological studies of activity are of fundamental nature; they are the source of difficulties in those areas where that category is used as an instrument of analysis. Labour is the basic and the highest form of human activity, whatever aspects of activity might be stressed and whatever features of activity might be pushed into the foreground. Since labour processes are the substance of technics, attempts to interpret the phenomenon of technics (in particular, to give a definition of it) in terms of other concepts, such as the concept of labour function, appear to be well-founded and promising. The concept of labour function is effectively used by Soviet researchers in the description of the labour process. This attempt is all the more well-founded as works on the history and theory of technics often concretise goal-directed activity in terms of labour functions (61, 27; 94, 9).

Five functions are usually distinguished in the labour process in the literature on the philosophical-methodological problems of technics: the technological function proper, the transport function, the energy function, the controlling and the logical function (61; 102, etc.).[2]

The identification of the principal functions in the complex process of human labour has a considerable heuristic potential. In particular, it permits a better understanding of the purpose of technology and a sufficiently consistent classification of it. The concretisation, in terms of these functions, of goal-directed activity in labour clarifies the essence of the nodal elements in the development of technics, of the man-machine system. Thus the present-day scientific and technological revolution (of which the nucleus is automation and cybernetisation of production), which is essentially the process of handing over to machines certain logical functions pertaining to human mental labour, frees man from performing various functions in the management of production. Just as the 18th-century industrial revolution freed the human hand, passing some technological functions over to machines, so does the modern scientific and technological revolution free man's mind by handing over certain logical and management functions to cybernetical machines and devices, to automata. The development of technics, being a process of gradually freeing man from direct performance of labour functions, results in the fact that direct contacts between man and the object of labour are replaced by contacts mediated by technology, by artificial means of labour.

In view of the existing tradition of methodological analysis of labour and technics in terms of the concept of labour function it would be interesting to apply this concept to the description of simple elements of labour. In other words, we can attempt a consideration of human labour in its abstract form as a function.

The concept of labour (or production) function may be arrived at through generalisation of the principal five (or six) functions. The abstraction thus obtained will record the essence of the concrete labour functions—not only the five better known functions but also others belonging to other possible classifications. The abstraction of labour function in general affords, in our view, a better understanding of the nature of labour as such (Marx), a clarification of its relationship with activity and, most importantly, with technology.

Mathematically, a function is an operation, a process or transformation, the application of which to one domain of objects (the domain of a function) produces another domain of objects (the range of values) (128, 15, 16). Robert Stoll also uses the expression y is "the element into which f *carries* x" (186, 37, 38).

The mathematical concept of function is entirely abstract, but it also records, exactly and profoundly, the essence of functional relations in any sphere of nature or human activity. Proceeding from this concept of function and from Marx's familiar scheme of labour, labour itself may be said to be a function of which the domain is the objects of labour and the range of values, the products of labour. Labour as a function applied to the object of labour (the argument) yields the product of labour (the value). In other words, labour proper proves to be the activity of transforming objects of labour into products of labour.

The material, objective implementation of this function, its realisation and carrier are the instruments of labour which man places between the object and the production of labour. The instruments of labour as a materialised function yield products of labour (or its values) when applied to the object of labour (the argument). We know that labour as such begins with the production of the implements or instruments of labour. In other words, labour emerges as the objectification of labour functions in a definite form—in the form of artificial instruments of labour. Labour as a function is realised in the instruments of labour. Thus concrete artificial instruments of labour are objectified labour functions, while the artificial instruments of labour in general, taken as a whole, represent the objectified function of labour in general.

Since artificial instruments of labour are at least part of technics (production technics), the latter can be defined as objectified labour functions. Going a step further, we may say that technics comprises artificial material systems realising or performing man's labour functions. The narrowing down of the concept of technics to instruments of labour only, or its expansion to embrace all means of purposive activity in general are both debatable; in view of this, the following definitions of technics may be of some interest: (a) technics covers artificial material systems realising man's goals; (b) technics embraces artificial material systems performing man's functions in the whole of his activity.

Analysis of the conceptions of technics and the definitions of technics based on these concepts shows that researchers identifying technics with instruments of labour or instruments of activity in general proceed, as a rule, from apparently obvious and intuitively clear characteristics of labour, production, practice, and purposive activity. This obviousness and clarity are, in fact, illusory. The study of the relation between the category of labour and other concepts shows that this category is not rigorously defined in the literature, it is marked by great polysemy and is used in different meanings. Ambiguous and non-rigorous interpretation of these categories often results in contradictions and at times in nearly paradoxical situations and evaluations in studies in technics.

The conception of technics as instruments of activity contains a number of moot points and elementary inconsistencies most of which are connected with the concept of goal-directed activity. On the one hand, technics is defined as instruments of purposive activity, which implies that its definition as instrument of labour is inadequate. On the other hand (as was pointed out above), goal-directed activity is concretised in terms of labour or production functions. Neither is it clear, in view of Marx's identification of purposive activity and labour as such, why technics—the artificial instruments of purposive activity—cannot be regarded at the same time as instruments of labour. Furthermore, the concept of labour function is effectively used in studies in the history and theory of technics. In terms of these functions, technics is interpreted as objectified labour functions, which is equivalent, in our view, to its definition as instruments of labour.

An unambiguous definition of technics cannot at present be given, as there are too many debatable points in the elaboration of the concept of goal-directed activity, in the specification of science and of other forms of activity as forms of labour. There is no consensus regarding the nature of technics. Does it comprise all the instruments of labour or all the artificial means of purposive activity in general? The important point, however, is this: is technics as an ensemble of instruments—whether of purposive activity in general or of labour activity in particular—the subject-matter of technical knowledge, or is it not? In our view, this question must be answered in the affirmative, and further methodological, philosophical-epistemological studies in the subject-matter of technical knowledge must be directed towards the study of the interconnections between technics and technical knowledge.

The above analysis of the concept of technics, despite a certain vagueness of its definition, warrants important conclusions bearing on epistemological investigation of technical knowledge. In particular, technics must obviously be interpreted as an instrument of activity rather than as an object or an ensemble of objects. It does not matter much which type of activity is meant—labour activity or activity in general. Technics as an instrument is included in the process of activity, and technical knowledge is objectified, transformed into technics precisely in the process of purposive activity. The system of purposive activity naturally makes a decisive impact on the processes involving technics as its element, including the process of materialisation of knowledge. Purposive activity is the substance in which technical knowledge is engendered.

This conception of technics as an instrument of activity creates the necessary premises for reconstructing the genesis and for understanding the development of technical knowledge. Of special importance in the reconstruction of the formation of technical knowledge is the study of the moment of its origin. Correct understanding of this moment is extremely important for the clarification of the subject-matter of technical knowledge and of the correlation between technical knowledge and objective reality. As for the analysis of the mechanisms of the birth of technical knowledge, it is linked with the study of the initial structures of technics.

The initial structures of technics and of technical knowledge. For all these reasons, and in view of the obvious ambivalence of technics (as an ensemble of the instruments of labour and of instruments of purposive activity in general), it is necessary to consider the origin and the initial stages of the development of technics and technical knowledge in the context of the formation of purposive activity. This is all the more necessary as Marx defines labour as such in terms of purposive activity. On this approach, of special interest is, of course, the specificity of purposive behaviour of animals, in particular of Primates, of their instrumental actions, etc.

The mechanisms of the emergence of technical knowledge and its materialisation in technics, especially at the initial stages, have not been studied adequately. Extending as they do, though, across man's history and pre-history, they are amenable to description. Parallel with the transformation of the ape's brain into the human brain, knowledge was objectified or materialised in instruments of labour, in technology. Analysis of the materialisation of knowledge in the initial, the very first technical structures is important for understanding not only the history of technology and technical knowledge but also for the theory of these questions.

At present, the mechanisms of the formation and realisation of goals are studied in cybernetics, philosophy, psychology, and in a number of other sciences. From the philosophical and psychological viewpoints, a goal is a realised image, an anticipation of future results. It has been noted in recent psychological literature that the problem of goal-setting and goal-formation has not been adequately studied. This is also true of the problem of purposive animal behaviour. This behaviour must not be anthropomorphised, of course, but neither is it possible to deny certain elements of intelligent activity, of the formation of notions and conceptual reflection in animals.

Considering the forms of psychical reflection that may appear as goals and levels at which the processes of goal-setting and goal-formation may unfold (74; 75; 95), the question may be asked: why cannot animals, capable of the formation of general ideas, of initial forms of conceptual reflection and even pre-verbal thinking, have such specific ideas or at least perceptual images as goals? If we recognise "embryonic" forms of labour in animals (Marx), we should also recognise embryonic forms of goals among them, for labour as such is goal-directed activity.

Scientists studying animals' purposive behaviour, more and more tend to think that they have certain "inner correlates" (146) or notions about goals (155). These conclusions, based on the analysis of vast empirical materials and fine experiments, confirm the proposition concerning the conscious, planned character of the actions of animals, expressed by Engels in his *The Part Played by Labour in the Transition from Ape to Man.* The assumption of the existence of a certain inner correlate of the goal-setting situation in animals' purposive behaviour, along with the striking examples of building activity and nearly proved existence of elements of intelligent activity in animals (46) suggest the idea that technics, at least in its embryonic forms, can exist among animals. The first step in the study of this hypothesis is analysis of the mutual relations of purposive behaviour and instrumental activity among animals.

K.E. Fabri writes that "the generally accepted view is that animals' instruments are external objects, i.e., objects that are not within the organism, not a part of the animal's body (not pertaining to its morphology), which act as auxiliary means and increase to some extent the effectiveness of behaviour in some sphere of life activity or even the level of behaviour as a whole" (105, 8). It is extremely important that in using an object as an instrument, the animal must be in direct physical contact with it, establishing a link between two objects—the implement and the object of action.

As far as animals' instrumental activity is concerned, it must be stressed that it has an adaptive, a purely biological function and is always determined by ecological factors, subject to general biological laws. Animals' instruments play an auxiliary role and are used accidentally, often in extreme situations only. The biological nature of instrumental activity is reflected, among other things, in its causes. It comes into play under the influence of hormones, in the presence of another individual, during pregnancy, etc.

Instrumental actions and purposive behaviour in animals (in the higher Primates) are not systematically connected—they are even separated from each other. The situation was apparently the same for man's ape-like ancestors, although their instrumental actions must have been developed better than modern apes'. The links between animals' purposive behaviour and instrumental actions is accidental, their basis is as a rule biological. This is confirmed by the familiar use by anthropoid apes of man-made implements for the biological purpose of domination.

Are instruments of action the means of attaining goals in these cases? In other words, are animals guided in these situations by some idea of a goal? That is apparently the case under experimental conditions, from which the conclusion may be drawn that under natural conditions the situation is the same. In both cases, however, the links between goals and instruments are accidental, even among higher Primates, as shown by observations under experimental and natural conditions. Thus technics as interpenetration of instrumental and purposive activities, as their systematic and necessary connection, as an ensemble of artificial means of purposive activity, does not exist among animals.

Instrumental actions and purposive behaviour reach a new level of development when they are systematically combined. The systematicity of this connection as a stage in their development was a consequence of tools and purposive activity becoming necessary conditions of some Primates' survival (103). This necessity was in its turn conditioned by an objective factor—by the fact that man's ancestors, the Primates, moved to the savanna.

In the struggle for survival that resulted from this move, instrumental actions became the principal, if not the only, and therefore the necessary means of achieving goals—satisfying hunger, protecting life activity, etc. The transformation of instrumental activity into purposive activity, that is, the merging of these two types of activity, was gradual. This interpenetration of instrumental and purposive activities brought about the time when the achievement of a goal was mediated by a tool made with the help of another instrument of action. At this stage, labour activity emerged.

Labour begins with making an instrument of action with the help of another instrument of action. In specifying the initial forms of labour activity, instruments of action should be understood precisely in the sense indicated above, so that the effectors of apes, for instance, are not seen as instruments of action. The formal premise of the emergence of labour—combining instrumental activity with purposive activity—occurs irregularly in the animal kingdom in the process of satisfying the need for food, shelter, etc. Such a synthesis, though, is also possible on the biological, instinctual basis, i.e., in the form of instinctive labour characterised by biological motivation, the pressure of need at the moment of the "labour" action. We have human labour when needs are satisfied through changing the form of a natural substance with the aid of a tool which is in its turn altered by means of another instrument of action.

The emphasis on changes in the form or structure of a natural substance is of prime importance for understanding the initial stages of labour activity. This aspect of labour is also of great significance for highly developed forms of labour, as indicated on numerous occasions by Marx (59d, 174, 177). The special significance of this aspect for the interpretation of the inception of labour lies in the fact that in satisfying his needs primitive man processed and altered those aspects or strata of natural objects that are described in terms of form or structure. One may say, in a sense, that the object of man's labour activity was the external aspect of natural objects, that which was intelligible and accessible to the senses equipped with as yet undeveloped thinking. The tools of labour activity—the instruments of man's action upon objects—corresponded to the goals of the initial phases of this activity. Primitive man's instruments of labour, his technics, were directed towards changing the form of natural objects, in the first place through division of the whole into parts, through combining wholes or parts, mixing particles of substances, etc. (89, 181). This shows that the technological function was, logically and genetically, the first function of labour.

The question of the first instruments of labour is highly important and essential in the context of studies in the initial forms of technical knowledge, of the initial structures of technics. We refer here precisely to the first instruments, i.e., a definite type of labour implements. Finding and describing the very first, the only instrument of labour is a scientifically insoluble task. The answer to the question about the first type of labour implements is at the same time the solution of the problem of the most ancient form of primitive man, as the first creators of artificial labour instruments were the first men. There are different views of the complex problems of the inception of labour activity and of man; of these, the most well-founded theory is, in our opinion, the view that labour, that is, systematic production of the first implements of labour, first came into being in the Oldovan culture of the *Homo habilis*, the most ancient member of the human race (67, 6, 29; 104, 99, etc.).[3] Labour implements appeared among the *Homo habilis* simultaneously in various forms—choppers, toothed and other types of implements. Of these, choppers were the main type.

The view that primitive labour implements are objectified knowledge is shared by most, if not all, scientists studying this problem (126; 130; 155; etc.). Moreover, V. Gordon Childe insists that the primitive labour implement materialised a concept, not just knowledge (125). Several questions may be asked in this connection. What place does knowledge materialised in primitive labour implements occupy in the structure of relations between purposive activity and labour implements? What part of these two simple elements does this knowledge reflect? Finally, in what forms does such knowledge exist?

In analysing the typical technological situation of the primitive man—working in stone—we notice the following. Working in stone, man apparently accumulated knowledge of its properties—weight, size, outline, etc., but it was not knowledge of these properties that turned stone into a labour implement. In lending stone a definite form, primitive man foresaw the situation of using it, i.e., he bore in mind its function. He made stone conform to a goal—the object of his need. These objects or goals were the animals he hunted, or the roots he gathered. The shaping of stone to suit a purpose was also based on past experience—observation of the action of stones under natural conditions, experience in using slightly altered natural implements, and so on.

Conformity, adequacy, fitness for a goal of purposive activity aimed at an object of some need was a most important element of producing a labour implement. In this process, the goal existed in primitive man's consciousness as a guideline for technology. In creating a concrete labour implement, man unfolded, as it were, the process of its application in his consciousness, he foresaw the situation of action, of contact between the implement and the object of his need, i.e., his goal. The principal formative element of the first stone implements was knowledge of situations of their employment, knowledge of the mechanical and physical properties of the goal, knowledge of the ways in which the mechanical properties of stone given suitable form act on the object, etc. Primitive labour implements thus embodied knowledge of purposive activity involving these implements.

From the epistemological viewpoint, of considerable interest is the question of the methods and forms of technical knowledge at the initial stages of its existence. Among the cognitive abilities necessary for the making of the first labour implements one must first of all point to memory, without which analogy, transference of experience and ability for imitation are all impossible (155, 39). Among methods, rudimentary forms of observation, analysis and synthesis must be mentioned in the first place.

Comparing the data of ethnography, archaeology, psychology, the history of technology, it is possible to trace the mechanisms of the work of cognitive structures, the application of the methods of technological knowledge in primitive labour implements. How do these methods and structures work, say, in making a trap—one of the first automatic devices?

Just as all the other stationary mechanical hunting implements employed by primitive man, traps used the weight and flexibility of bodies. Primitive man did not of course know the foundations of physics and mechanics, but he could see animals—objects of his need—become accessible to this need through falling into a natural hole, through being struck down by a falling tree or stone, etc. Thus man first observed the mechanisms of combination of goals and instruments for their achievement in nature. Moved by his needs, he began to use, first of all, the cognitive arsenal inherited from animal ancestors—imitation, ability for situations transference. That was why, imitating natural situations, man at first drove animals into natural holes, bogs, etc. At the next stage, continuing to imitate nature and at the same time realising a situation transference, man changed natural forms, digging holes, felling trees, etc., constructing in this way situations for the attainment of goals. For example, observing animals who died through getting caught among lianas, primitive man constructed snares on the same principle (163, 76-92).

The process of constructing the first self-acting devices developed from mechanical imitation to creative changes of the natural factors on the basis of the analogy principle. We find evidence of creative imitation in constructing the first labour implements not only in the development of primitive single-action automatic devices, such as a noose, but also in combining them with various types of fences directing the movement of animals.

Primitive technical structures embody knowledge of two types: knowledge of the functional and morphological characteristics of the implements of labour; knowledge of the function or purpose of a labour implement and knowledge of the natural material of which it is made. The main thing that makes a labour implement what it is is the knowledge of its function, the purpose for which it is created. Apparently knowledge materialised in the primitive technical structures is a reflection of goal-setting activity, of the functions of labour implements. Knowledge of morphological, material characteristics of labour implements is conditioned by the knowledge of function.

Goal-directed activity as an objective process is an object independent of technical knowledge and reflected by this knowledge in primitive technical structures. Knowledge correlated with primitive technical structures has a complex structure, consisting of at least two parts. It reflects purposive activity materialised in labour implements, in other words, their functions, and also the properties—mechanical, physical, chemical, etc.—of the material. Thus primitive technical structures reflect (and we mean precisely reflection, not correlation of an indefinite nature) goal-setting activity and nature as objective processes. Knowledge of goal-setting is here the main thing, an instrument of purposive change of natural objects.

The structure of the subject-matter of technical knowledge. Analysis of the objectification of goal-directed activity at its early stages, the study of the isolation of labour activity proper from purposive activity in general, and the evolution of technics is of great importance for epistemological inquiry into technical knowledge. Having reified purposive activity (the process of goal-setting) in objects, having thus made the means of achieving goals artificial, man later turned, in studying purposive actions (labour functions), to already created object structures. The impression is therefore created that only these object structures are the object of technical knowledge, the more so that they are sensuously perceived and largely objective.

At the early stages of purposive activity the goal is the object of need. It forms and determines the means for its attainment. The subject (man), the vehicle of the goal, objectifies it and forms the object as a means of achieving the goal, and performs this with the aid of the knowledge of the goal, its interrelations with the means of achieving it. Man makes the object, as a means, conformable to the goal, i.e., purposive. The activity to form the object in accordance with the goal, purposive activity, "wanes" in the means intended to achieve the goal.

Satisfying the immediate physical need and the transition, according to Marx, to free production brings about a situation in which the need is satisfied through many objects. As far as goal-setting is concerned, this is expressed in the fact that goal as reflection of a given concrete object shaping labour implements gives way to task and later to function. Implements of labour are separated from goals as the concrete objects of need. In other words, the goal shaping instruments of labour, the technological structure, assumes in the development of technology a different status, becoming a task or a function.

Gradually, technics begins to improve with regard to tasks and functions, regardless of the concrete goals and needs connected with the achievement of a given object. In his creative technological work, man now turns to purposive material structures. But the connection between technical structure and goal is not lost. It is present in the explicit or implicit realisation of the function determining the material purposive structure in the solution of technological tasks. This tendency in the development of technics including, in particular, the transformation of goal into task and function, began in Stone Age technology and has continued to advance up to the present. A classical example of the improvement of a task—or function—oriented technical structure was the invention of the steam-engine (44, 23-24).

The programmatic-purposive approach to the construction of major technological projects and generally to material and nonmaterial production, which has sharply gained in strength in recent years, created the premises for the establishment of the decisive role of goal in man's practical activity, including engineering and technology. The formative character of goal in relation to technical structures is strikingly manifested in "non-classical" forms of engineering and technical knowledge. Construction of complex engineering projects has brought to light the purposive nature of the functions of technological structures. That is one of the principal elements in the methodology of engineering design, ergonomics, systems design, systems engineering, etc.[4]

Research in the methodological problems of "non-classical" forms of engineering activity and of scientific-technical disciplines is at present in its initial stage. The forms of organisation of "non-classical" technical knowledge, the interrelations between disciplines, and the relationship between the knowledge and engineering activity are not yet fully realised. In particular, there is no clarity so far about the relations between systems engineering, ergonomics, and systems design, or between ergonomics and human engineering, etc.

And yet, methodological analysis of creative elaboration and introduction into practice of complex systems at the present stage, and generally the study of the processes of creating technology throughout history shows that goals determine the development of technological structures. Philosophical-epistemological and psychological studies of task, problem and design in creative technological thinking also point to goal as the core of engineering and technological activity (70; 78). Russell Ackoff, the well-known specialist in the methodology of problem-solving, also connects solution of problems with achievement of goals (116, 19). Technical knowledge embodied in technical structures is a reflection of man's purposive activity as an objective process. That explains, among other things, the fact that engineering-technical knowledge takes the form of instructions.

Of the greatest interest for our research is the question of the nature of man's goal-setting activity, of purposive transformation of society and nature. Are they the object or the subject-matter of technical knowledge? We believe that man's goal-setting activity as a form of the objective process (51, 38, 188) is the object of technical knowledge. The objectiveness of goal-setting activity, i.e., its independence from consciousness is a consequence of the dependence of goals on nature, on natural laws. Goal-setting activity is objective in content, in its dependence on the needs, and in its results.

Man's goal-setting activity is not, of course, objective in the same sense as nature. There are two forms of the objective process. Man's goal-setting activity apparently includes subjective factors as well. A goal may exist in different forms of the psyche, of consciousness, and goal-setting may take place at different psychical levels. Besides, the activity of goal-setting and achievement of goals includes a purely psychological ingredient (74, 157-167). This last point is of great importance for understanding the mechanisms of implementation of knowledge about goals in technological structures, for understanding purposive activity as the object of cognition. It is in the sphere of the psychological constituent of purposive activity (of the result of this activity) that technical knowledge is formed—at the initial stage, at any rate.

The subject-matter of technical knowledge is materialised purposive activity, i.e., technics in the above sense—artificial material systems realising man's labour functions (goals). As we have pointed out, the term "artificial" in this definition is synonymous with "man-made".

The problem of the artificial is the key problem as far as technics is concerned. The fundamental nature of this problem explains certain difficulties in its solution—in the analysis, for instance, of the first stone labour implements and of their difference from stones "processed" by nature. The idea about close links between goals and the artificial seems to be one of the most reasonable and well-founded ideas concerning the general approach to this problem (183, 6-7). These close links permit, in fact, the transition from the definition of technics as a means of purposive activity to its definition as artificial means of activity, artificial systems, etc. Previously considered definitions of technics despite all differences between them, have a common objective content, they pertain to one and the same class of phenomena.

It has been pointed out above that, in the framework of the activity-oriented approach to technical knowledge, technics can be interpreted as the subject-matter of technical knowledge. To the arguments adduced in the above, we can add here the following. The artificial can be described, as in (111), in terms of structure and function. But the artificial can also be described, in our view, from the standpoint of goal-directedness. The artificial is then a synonym of the goal-directed.

Taking into account the purposive nature of function, which in its turn determines the structure, the morphological characteristics of the technological object, we can conclude that the functional-morphological description of the artificial and its purposive interpretation are two equivalent systems of concepts characterising one and the same subject-matter—technics. This situation is analogous to the relationship between ergonomics and human engineering, which have an identical subject-matter but different organisations of knowledge. Therefore, the view of the artificial as the subject-matter of technical knowledge (or sciences) (35, 220) can be interpreted as a reference to technics in the above sense (the subject-matter of technical knowledge).

Considering the ambivalence of technics and the assessment of this ambivalence in connection with the epistemological analysis of the subject-matter of technical knowledge discussed above, we believe that the definition of the subject-matter of technical knowledge as the totality of artificial material systems to be the most adequate view. The question of whether these systems are implements of labour or of purposive activity in general here recedes into the background, it becomes insignificant. The view of technics as the subject-matter of technical knowledge is in complete agreement with the most general criterion, evolved in epistemology, in the methodology of scientific cognition, of science (or knowledge) as a unity of the subjective and the objective, technics being materialised purposive activity, an embodiment of technical knowledge.

The two-layer structure of the subject-matter of technical knowledge in this interpretation is obvious. The two-layer structure of the subject-matter of technical knowledge determines the two-layer character of knowledge in non-classical technical sciences such as ergonomics, systems engineering, etc. Concretisation of the structure of the subject-matter of technical knowledge, the identification in it of goal-setting activity facilitates effective research not only into the complex problems of the organisation of knowledge in non-classical scientific-technical disciplines, the genesis of the primitive structures of technics and technical knowledge, etc. We believe that on the general philosophical, epistemological plane, the approach to the subject-matter of technical knowledge developed here creates the premises for an explanation, more concrete than other existing explanations, of the content of technical knowledge in its pre-scientific, classical and non-classical forms, and of the relation of this content to objective reality. On this approach, certain essential aspects can be traced of the reflection in technical knowledge of its object and subject-matter.

The view of the subject-matter of technical knowledge as artificial material systems, as objectified goal-setting activity, enables us to take a more advantageous approach, supported by actual engineering and technical practice, to the analysis of the conceptions of the subject-matter of technical knowledge in contemporary foreign literature, especially British and American. In evaluating the methodological, epistemological interpretations of technological knowledge (and of technology) by Mario Bunge, James Feibleman, Henryk Skolimowski, I. C. Jarvie, Stanley Carpenter, and others, we must point out the actual aspects of technological knowledge and the rational elements in its interpretation which were raised to an absolute and exaggerated in their theories.

Analysis of the works of these authors, collected, e.g., in the well-known book (172), shows that they share a number of common theses. The first of these is, of course, the correlation of technological knowledge and technology, just as technical knowledge is correlated with technics. "Technology" is a polysemantic word, so that in some cases the relation of technological knowledge and technology assumes the quality of identity, technology being regarded as a type of knowledge. It is obvious, however, that the two senses of the term "technology" must be differentiated; this is clearly realised by those scientists who closely study the problem. Thus Carl Mitcham believes that the term "technology" has four senses: (1) technology as object; (2) technology as process; (3) technology as knowledge, and (4) technology as volition (171, 306). Technology as knowledge (technological knowledge) is connected in the first place with technology as process, which is especially stressed by Carpenter (124, 162).[5]

The common feature of various interpretations of technology as process, ranging from Feibleman's skill to Jarvie's know-how, to Bunge's technological theories, is technological action to which technological knowledge is related. The clearest, best thought-out and logically accomplished theory of technological action has been suggested, in our view, by Eugen Olszewski. His system of propositions on this problem may be said to have deduced all the conclusions from the principles describing technological actions.

Olszewski's complex typology of human actions (175) consistently identifies practical, economic, and, within the latter, technological actions, which have the following characteristics. As practical actions, they change reality by means of all kinds of innovations. As practical economic actions, technological actions are consciously realised and goal-oriented. Among other economic actions, technological actions proper are marked by the fact that all or most of the means of achieving goals are man-made or inorganic in nature. Thus if man sets himself a goal and achieves it with the aid of some tools he made himself (these goals and tools being inorganic in nature), this action may be described as technological. Technological actions are different, e.g., from medical actions, for, although the tools are usually man-made, the goals are connected with natural objects and processes, i.e., with organisms and biological processes in man and animals. The world of technology, i.e., all technological actions and their results, is, according to Olszewski, the subject-matter of technological sciences, of technological knowledge.

The conceptions of technological knowledge discussed here have a number of defects pointed out both in the Soviet and Western literature on epistemological problems of technological knowledge, such as the absence of any analysis of the structure of technological knowledge, of practically any studies in the genesis and development of technological knowledge, neglect for or, to be more precise, relegation into the background of substantive, objective aspects of technology, etc. But the views of these authors contain one element which is extremely important for defining the specificity of the subject-matter of technological knowledge, namely the view that technological actions are that which technological knowledge is directed towards, starts out from, and closes upon. In the conceptions analysed here, this aspect of the subject-matter of technological knowledge is hypertrophied at the expense of the substantive aspects of technology. In substantiating this proposition, one aspect of the situation is raised to an absolute, and that, of course, is an error. The rational kernel here is the link between technological action and man's goal-setting activity, which explicitly figures in Carpenter and Olszewski.

This conception of the subject-matter and object of technological knowledge permits a new view of certain complex problems in the analysis of the subject-matter of "non-classical" forms of scientific technical knowledge—systems engineering, ergonomics, systems design, and other systems-oriented technological disciplines. For example, what is the objective content of the notion of integral system in the construction of complex engineering objects? Similar questions can be raised in the context of ergonomics or systems design. The possible answer to this question is closely linked to the study of the synthesis of knowledge in non-classical technical sciences, which we shall discuss below. Right now we are interested in another question—what conditions the integral quality of the notion of a complex object that is being designed and does not yet exist in objective reality. In technological practice, this task is solved in each particular case with the aid of simulation (22, 66-67). From the epistemological standpoint, the objective phenomenon ensuring the integrality of the design image is the goal as a link in goal-setting activity which is a form of the objective process. One of the aspects of the formative character of goals in relation to technological structures, is, as was pointed out above, the fact that goals determine the integrality of the conception of a complex engineering object at the stage of design.

The status and structure of technical knowledge. Defining the subject-matter and the objective sources of technical knowledge is only one, if highly important, element in its epistemological characterisation. To give a more comprehensive methodological description of its scientific and prescientific forms, we must turn to the content, structure and functioning of technical knowledge. Technical knowledge is knowledge materialised or objectified in technics, knowledge about the man—technics (or man—machine) system; this became obvious with the emergence of non-classical scientific technical disciplines. Epistemological studies in classical technical sciences implicitly reflected this fact in correlating technical knowledge with technics defined as a goal-directed structure.

Two aspects or groups of problems can be identified in epistemological (methodological) studies of technical knowledge; solution of these two groups of problems yields two groups of its epistemological features or characteristics—one pertaining to the status of technical knowledge and the other, to its structure. In the first case, technical knowledge is regarded as a system, an integrative property that cannot be reduced to its own elements—theories, hypotheses, various schemata or separate disciplines. Questions of the status of technical knowledge are concerned with its place in the structure of scienific knowledge in general, the study of the correlation between technical knowledge and natural knowledge, mathematics, the humanities, etc. Here also belong the problems of the subject-matter and emergence of technical knowledge, the study of the methods, ideals, norms of organisation and, generally speaking, analysis of the external functioning, as it were, of technical knowledge in the structure of scientific knowledge in general.

Problems of the structure of technical knowledge pertain to the study of the proper elements of technical knowledge forming its system. On this plane, technical knowledge is considered from within, as it were, without any correlations being established between its elements and other forms of knowledge, scientific and prescientific. A typical case of the analysis of the structure of technological knowledge is inquiry into the formation and structure of theory in the technical sciences, which implies studies into the nature of ideal objects, technological hypotheses and facts—generally speaking, of problems in the empirical and theoretical levels of technical cognition.

The status of technical knowledge involves problems that cannot be consistently considered outside their links with other forms of knowledge, outside the context of the principal kinds of scientific cognition and culture in general. At the same time there are questions which can be called problems in the structure of technical knowledge and which can in principle be studied in the framework of technical knowledge. This division is fairly relative in character and is closely linked with the depth of research into the status and structure of technical knowledge. Of course, consistent analysis of the specificity of technical knowledge, comparison of this knowledge, e.g., with physico-mathematical natural science, necessarily leads to studies in the elements of technical knowledge, and vice versa. In particular, research into the structure of technical theory implies, at definite stages, appeal to the results of analysis of the structure of physical and mathematical knowledge, its theories, etc. We can therefore say that identification of these two groups of problems is, to a considerable extent, the choice of the plane on which technical knowledge is considered, which does not change, of course, the nature of technical knowledge.

This classification of problems reflects the main direction of the efforts of methodologists studying technical science. Most Soviet and foreign publications deal with the problem of the status of technical knowledge, and, above all, with its subject-matter and object. The structure of technical knowledge is studied less closely. Taking into account the inalienable links between the status and structure of technical knowledge, we can assume that the problem of the epistemological status of technical knowledge has not been solved because of insufficient attention to the problem of its structure.

Technical and applied sciences. An essential element of the epistemological analysis of the status of technical knowledge is its relation to applied knowledge. The absence of complete clarity on this issue or, at least, of generally accepted concepts is determined not only by the complexity of the interrelations between technical and applied studies. The difficulties also arise here due to a certain relativity and vagueness of the division of sciences into fundamental or pure and applied. In other words, clarification of the interrelations of applied and technological knowledge requires a clear understanding of the entire triad: fundamental sciences—applied sciences—technical sciences.

This schema is fairly widely current among Soviet and foreign authors. The main thing here is the content we ascribe to this triad, the way we understand the separate elements of the chain, and the links between the components of this structure. In attempting to solve these questions, some Soviet scientists support the following view. The "fundamental sciences—applied sciences—technical sciences" sequence is a concretisation of scientific knowledge. Concretisation in this context is taken to mean not only the transition from higher-level abstractions to lower-level abstractions but also the exclusion of higher-level abstract objects in the application of theory to practice (107, 51).

It is this scheme that is, in fact, analysed by Feibleman (172, 33-41). Distinguishing between pure and applied aspects in science on the criterion of goals, Feibleman describes technology, as we have previously remarked, as skill. Technology is in his view a further approximation of practical tasks by applied science, being, so to speak, a scientifically substantiated improvement of the instruments of activity. In evaluating Feibleman's position on this issue, we must say that he fails to notice the production of new knowledge in the approximation of practice by science. Applied science is for him pure science applied to practical problems. Similarly, technology is a further application of the same pure science—for different purposes. Feibleman ignores the fact that the point of view, motives and goals do not change the nature or content of the object under study—including scientific knowledge. That is why both technology and applied science, considered on the content plane, are nothing but pure science.

Mario Bunge views the relationship between fundamental and applied science from similar positions, using motives as the feature distinguishing fundamental studies from applied ones. Bunge's conception of technological knowledge is one of the most widespread and widely recognised theories abroad, primarily in the English-speaking countries. Its analysis is therefore of special interest and significance. According to Bunge, applied science is pure science directed towards the solution of practical tasks. He views technology as actions and studies oriented towards changing reality—natural or social. Differences between tasks naturally produce different technologies. Bunge's works (see, e.g. (122)) provide a fairly detailed classification of technologies. In Bunge's view, technology is essentially an applied science, although Bunge himself speaks of certain nuances distinguishing them, without going into the details of these differences.

Bunge does not analyse the structure of technological knowledge or, more precisely, the structure of technological theory; he therefore does not see the new quality distinguishing technological knowledge from pure science. He identifies technology with applied science, i.e., pure science applied to practical tasks. Like Feibleman, Bunge does not notice that application of pure science does not only provide solutions for practical problems but also yields new knowledge different from pure science. He only registers the fact itself of the application of pure science and its result, losing sight of the birth of new knowledge in the process.

The examples cited by Bunge to illustrate his viewpoint can be used to show the erroneousness of his position. He writes, for instance, that "a theory of flight is essentially an application of fluid dynamics" (172, 63). An analysis of the mechanism of this application of fluid dynamics and of the result of such application—flight theory (169) shows that this application results in new knowledge that is absent in fluid dynamics.

We get similar results in studying Bunge's another example, that of psychology (ibid.). In considering the possibility of applying psychology to the solution of production tasks, Bunge actually ignores engineering psychology which is born in the process of this application and has a content of its own. It is very difficult to explain, from Bunge's positions, the differences between technological sciences emerging on the basis of electrodynamics, such as radio engineering and electrical engineering. Bunge's conceptual apparatus merely permits the statement that the two substantive technological theories are applications of electrodynamics.

Applied and technological sciences are sometimes identified on the basis of the existence and analysis of technological disciplines—whereas Bunge arrives at this identification via a study of pure and applied research.

The mutual links between applied and technical studies are not unambiguous. The study of their relations is a complex problem that has a number of debatable solutions. However, the identification of technological and applied sciences, whether it starts out from an analysis of applied studies or technological studies, is based either on vague criteria for differentiating between fundamental and applied knowledge or on an amorphous, undifferentiated view of practical tasks, or else on an inadequate perception of the mechanism of application of pure science to applied tasks.[6] These causes mostly act simultaneously. The situation is also made even more complicated by the rapid development of technological knowledge and the emergence of new discipline, including those of fundamental nature.

Methodological studies of scientific technical knowledge pay little attention to the specificity of the tasks solved by technical sciences, although that is one of their differences from applied science. Technical sciences do not simply solve practical or, say, industrial tasks—they concentrate on technics, on technological problems. This is especially stressed in (10), although the view that technical sciences have to do, above all, with technics, is fairly widespread, as pointed out in the previous section of the present chapter.

Among scientists concerned with problems of the history and methodology of technical sciences, the general tendency is to reject their identification with applied sciences. In his analysis of the problems of the history of technology and its modern state, Melvin Kranzberg writes: "We have also done away with the old maxim that technology is simply applied science..." (122, XIX). Symptomatic is also the fact that, whereas the article on science in the *Great Soviet Encyclopedia* (1974) identified technical and applied sciences, there is no such identification in (3).

The study of interactions between fundamental, applied and technical sciences is necessary not only from the methodological standpoint. This question is also of great significance for the constantly emerging new sciences, such as rock physics. A recent article on the methodological aspects of that science points out that "rock physics is a very young science, and its further development largely depends on the image of science which scientists active in that field take for a model of constructing scientific knowledge" (91, 333).

A clear realisation in methodology of the specificity of fundamental, applied and technical studies, and subsequent application of the specified characteristics of scientific knowledge to a concrete scientific discipline provides additional, and sometimes the only possible means and ways of solving theoretical and practical tasks of the development of the given discipline.

In the case of rock physics, its classification as a scientific-technical discipline rather than as an applied branch of solid-plate physics permits a clarification and definition of its sphere of application and a definition of its subject-matter. It is now methodologically justified to distinguish theoretical or fundamental research in it, not only applied. Methodological clarity concerning the interdisciplinary nature of this branch of science opens up further heuristic possibilities in the solution of real problems in rock physics such as synthesis of knowledge, specification of the complex instruments of cognition, etc. Problems of this type are often solved on the analogy of cognate problems from other interdisciplinary technological sciences. Furthermore, correct understanding of the actual scientific practice, reflected in the correct classification of the given discipline as fundamental, applied, or technological, creates the premises for the necessary restructuring of professional consciousness and the system of personnel training, the system of education. As a result of all this, the name of the science is changed to "theoretical geotechnology".

The inception and development of technical sciences. The "fundamental sciences—applied sciences—technical sciences" schema may and must be discussed as a sequence of stages in the concretisation of scientific knowledge. On this plane, however, it needs greater substantiation. The existing few interpretations of the concretisation of knowledge are open to criticism. The concretisation of abstractions in the movement of theory towards practice, far from cancelling abstract objects of a higher level at the stage of applied and technical sciences, fills them with new content. In the process, the essential characteristics of the fundamental models are not lost but are seen in terms of particular instances.

The analysis of applications of theoretical mechanics and of the development of the technical sciences connected with mechanics, which we find in A. Mandryka's books (57; 58), shows in detail, citing examples from various periods and theories, the difficult process of re-interpretation and concretisation of the fundamental concepts of theoretical mechanics in the specific laws of these technical sciences. The example of the technical sciences of this type is typical of the relations analysed here. The complex relationship between the electrodynamics of Maxwell and that of inventors and technicians ultimately resulted in the emergence of electrical engineering, in which all the practical concrete actions of the electrical engineer are based on Maxwell's basic equations. Similar tendencies are observed in the formation of other sciences based on electrodynamics—radio engineering, later radiolocation, etc.

Research in the actual development of scientific cognition suggests that the study of the "fundamental sciences—applied sciences—technical sciences" triad is of fundamental significance for the analysis of such an important problem of the status of technical knowledge as the inception and later development of technical sciences. Moreover, this schema may be said to be of prime interest for the methodology of science, reflecting as it does the essential elements of the development of technical sciences.

Before reaching the scientific stage, technical knowledge went through a number of forms in its development. In the above, we described in sufficient detail the inception of technical knowledge, its initial forms and methods. Accepting on the whole the schema for the formation and development of technical knowledge suggested in (35), we must clarify certain details and concretise a number of propositions—a natural move in the elaboration of any schema.[7]

Relying on the evidence of archaeology, anthropology, ethnography, etc., a study of the prescientific stage in the development of technological knowledge permits something more than just placing it in the period between the primitive communal structure and the Modern Times and partially structuring it, so to speak—it also permits approximate dating of the emergence of techological knowledge. The solution of this problem may appear to be trivial: we can get a dating, albeit approximate, of technical knowledge merely by pointing to the first tools in which technical knowledge was materialised. It appears, however, that the act of such pointing assumes the solution of extremely complicated problems of anthropogenesis, the structure of human labour, the artificial and the natural, etc.

The view was accepted in the previous section that the first type of labour implements were choppers. Despite differences in the first tools, repetition of the forms and similarities in working in stone point to the goal-directedness of the handling of these tools. The constancy of the forms of tools handed down from generation to generation indicates conscious purposive actions involving the first labour implements. Thus choppers and other tools of the Oldovan culture are a materialisation of goal-directed action; in other words, Oldovan tools are the first constant, systematic form of objectified technological knowledge.

In analysing the forms of prescientific technical knowledge, one should point out the fact that distinguishing in it the practical and the technological forms of technical knowledge is somewhat problematic. It is no accident that Carpenter in his study of the forms of technological knowledge uses one term for both these types of technological knowledge—the term "skill", making no practical distinction between them (124). Technological knowledge is born with the first stone tools, and it would be incorrect to view it as development and increase in the complexity of the knowledge of practical methods. Technological knowledge of the prescientific stage is, essentially, empirical knowledge of practical activity. It takes ages to accumulate this mixture of ignorance and practical skills by trial and error. Being part of folk wisdom, technical knowledge is at this stage a system of thinking based on common sense. This last point was noted by Alexander Koyré, but he raised it to an absolute, so that technology, according to Koyré, is a system of thinking based on common sense, absorbing certain elements of scientific knowledge and linked with intellectual history but independent of science.

The evolution of the scientific form of technical knowledge is connected with the transition to machine production. The development of material production and technics required engineering solutions of production tasks based on science, it required mathematical calculations in designing technological devices. Technology could no longer develop on the basis of mere common sense, keenness of wit, empirical experience. That was why the birth and formation of technical sciences, of which technics was the subject-matter, "was determined by two oppositely directed processes: on the one hand, by the use of natural scientific laws, theories, and separate data in the study of technological objects and the processes taking place in them, and also by active application of the methods of scientific cognition, and on the other hand, by generalisation of separate observations and facts of technology and production (102, 58).

Dissection of the resultant of these oppositely directed processes and the study of the nature of their synthesis reveals the fact that technical sciences were called to life by the needs of technological and engineering practice. The role of natural-scientific knowledge in the genesis of technical sciences cannot, of course, be ignored either. The practice of engineering was, so to speak, an external stimulus of the formation of technical sciences. But that does not mean that technical sciences have grown out of prescientific forms of technical knowledge by themselves. The birth of technical sciences was more the result of implantation of natural-scientific and mathematical knowledge in technological activity. It is therefore wrong both to ignore the role of natural-scientific knowledge in the emergence of the technical sciences and to interpret the latter as mere derivatives of fundamental natural-scientific theories. It has already been pointed out that the process of application of natural science to the technological problems of production gives rise to new knowledge irreducible to the knowledge of basic theory and the common sense of technology.

The hypothesis of derivation of technical sciences from natural science, which is regarded as independent from technical knowledge, appears especially doubtful in the context of the study of the birth of experimental science of the Modern Times. At the initial stages of its development, experimental natural science of the Modern Times started out from the ontology which emerged as a result of the objectification of prescientific forms of technical knowledge, i.e., of contemporary technics. The engineering tasks which stimulated the emergence of technical sciences also emerged on the basis of this technical ontology. But technics was also the domain which served as the "experiential basis for the emergence ... of the theoretical thinking of new physics" (4, 214).

In studying the emergence of scientific knowledge in the Modern Times, historians and methodologists of science note the influence of technics, of its spreading and improvement, on the social conditions and the mode of thinking. Technics (objectified technical knowledge), as a new ontology, made an impact on the worldview, laying the foundation of new culture. The social processes determined by the development of technics, and technics itself, naturally made an impact, too, on the thinking of major scientists, such as Galileo, who stood at the beginning of new science. Their theoretical thinking was inevitably affected by the influence of the technical ontology, and by the engineering, experimental tasks that it gave rise to. Moreover, in analysing the state of affairs in terms of derivation of some sciences from others, we see that we deal with the following, probably somewhat simplified, situation, as far as physics is concerned: "One might in fact say that new physics was born from an experimental branch of applied mechanics" (ibid.,215). In other words, physical-mathematical natural science emerged as a branch of technical science that was born at that same time.

The inception of technical sciences falls approximately in the period between the middle of the 15th century and the 1870s;

a characteristic feature of this period is the use of scientific knowledge for the solution of production industrial tasks, not just practical problems in general. During the first stage of this period (the second half of the 15th through the early 18th centuries), technical knowledge did not yet attain a theoretical level, since well-formed theories in natural science did not yet exist. This stage was marked by the formation of applied studies on the basis of experimental methods. The time between the early 18th and the late 19th century was decisive for the formation of technical sciences connected with physics, chemistry, and mechanics. The emergence of fundamental natural-scientific theories and well-developed technical practice created the necessary premises for the raising of technical knowledge to the theoretical level.

The aforementioned qualitative stages in the emergence of scientific-technical knowledge (fundamental sciences—applied sciences—technical sciences) can be placed within different chronological frameworks which, however, will not differ much. Thus A. Bogolyubov believes that technical sciences emerged by the middle of the 19th century, while the rise of applied sciences falls on the beginning of the 19th century (10, 81-82). It is also obvious that electrical engineering as a technical science came into being somewhat later than the 1870s—the period when Maxwell's famous equations of the electromagnetic field became more or less widely known. A similar situation existed in radio engineering. Importantly, this schema obtained in the technical sciences connected with physics, chemistry and mechanics, being, in fact, the schema of the inception of the technical sciences of the classical type.[8] One can therefore agree with such researchers as G. Ruzavin, who insists that "initially, technical sciences emerge for the solution of purely applied tasks on the basis of the results of such fundamental sciences as mechanics, hydraulics, physics, chemistry, etc. Later, the process of development grows more complex" (107, 53). Indeed, the development of scientific technical disciplines becomes much more complex, especially in non-classical disciplines.

But the mechanisms and the forms of the evolution of new technical sciences began to change significantly already in the "classical" period (late 19th-mid-20th centuries) of the development of technical knowledge. At this stage, the traditional mode of the emergence of technical sciences through derivation from the basic natural science continued to exist. As we have pointed out, derivation should be interpreted as synthesis of engineering-technological practice and natural-scientific theory.

In this way electrical and radio engineering were derived from electrodynamics. At the same time a new form of the inception of technical sciences came into being—through derivation from an already existing technical science, which in this case acts as a fundamental science. In this way radiolocation was, for instance, derived from radio engineering. In this case, too, the interests of technical practice, of engineering, form an essential aspect of the mechanism of derivation of a new scientific-technical discipline from an already existing technological science.

It should be noted that technical sciences in this period are already a well-formed field of scientific knowledge with its own subject-matter, its theoretical principles, and specific ideal objects. Original mathematical and conceptual apparatuses have already been worked out in a number of disciplines. The system of technical sciences assumes stable forms of relations with natural sciences. In this context, the separation of some technical sciences from others, as a new element in the mechanism of the emergence of the scientific-technical disciplines (when they are considered in a new aspect), is differentiation of technical knowledge in the precise sense. The latter is most characteristic of the "classical" stage in the development of technical knowledge. Another aspect distinguishing the classical stage from the previous stages of development of technical knowledge is the acceleration of the rate of mathematisation of technical disciplines, a qualitative leap in this process.

Differentiation of technical sciences at the classical stage, as e.g. the separation of the theory of machines, the theory of electric drive, etc., from theoretical electrical engineering, introduced new elements in the interpretation of the fundamental vs applied problem in its connections with technical knowledge. Differentiation and integration made possible and necessary the identification of general laws in the sphere of technological knowledge. The question thus arose of fundamental technical laws.

The problem of fundamental and applied sciences is especially acute in the methodology of technical sciences in view of their genesis, in view of the fact that in their evolution they go through the stage of applied knowledge. A. Bogolyubov rejects the existence of fundamental laws in the field of technical knowledge. In his view, technical sciences may become applied or even fundamental, but fundamental technical sciences do not exist. Ascending to abstract knowledge that does not close upon technics, technical sciences lose the quality and status of technical knowledge, becoming a different type of knowledge—applied or fundamental (10). Some authors speak of "applied technical sciences", transferring, in our view, the concepts and terms of natural science to a sphere where such connections have no place. The division of sciences into fundamental, particular, and concrete technical ones appears therefore to be the most correct schema (111, 133).

The formation in the classical period of a qualitatively new system of knowledge different, e.g., from natural science, stimulated attempts to define the specificity of technical sciences. Science may be regarded as well-formed if it effectively performs its principal functions—those of explanation and prediction. At the empirical level these functions—or at least prediction—are impossible. It follows that the question of well-formed science is a question of the formation of theory in it. Thus the criteria of the maturity of a science or, generally speaking, the criteria of its scientificity, are the criteria of the theoretical stage in its development.

This proposition is especially true of natural science. In technical knowledge, a science may be regarded as well-formed if it effectively performs its principal function—the constructive one, the function of generation of ideal engineering objects. The most effective instrument of achieving this is the mathematical apparatus, deductive mathematical reproduction. It is precisely in this sense that we can say that the technical sciences took shape in the 1920s-1940s, when mathematised theories were constructed in a number of classical technological disciplines which thus attained the second stage of mathematisation, at least (10).

Before the 20th century, the mathematisation of technical knowledge proceeded with difficulty; it only became stabilised in the 1920s-1940s, when the classical scientific-technical disciplines assumed their present-day form. The first sporadic attempts at defining the specificity of technical sciences in the methodology of scientific cognition early in this century were ineffectual, mostly due to the low theoretical level of technological disciplines in that period.

Technical knowledge became the object of serious epistemological analysis in the 1960s. At that time, the material of methodological studies was, as a rule, the classical technological sciences. The emphasis was on the specificity of technological knowledge compared to natural science, on establishing its domain, structure, etc. However, differentiation of technical sciences in the classical period, and the mushrooming of new technical disciplines at present, require greater concreteness of methodological studies in the field of technical knowledge.

At present, the most adequate unit of methodological analysis of scientific-technical knowledge is a scientific-technical discipline (91, 307). This approach permits not only the tracing of important social, organisational, communicative and other aspects of the development of scientific-technical knowledge—it also creates the premises for solving the important methodological tasks of studying theoretical schemas of separate scientific-technical disciplines, which is especially important for their accelerated development at a time of the scientific and technological revolution.

The need for the attainment of a new level of development in methodological studies of technical sciences, and in particular for focusing of attention on separate scientific-technical disciplines (the technical sciences) is determined by the emergence of interdisciplinary technological sciences. Complex technical sciences (technical sciences of the non-classical type), such as ergonomics, systems engineering, systems design, theoretical geotechnology, and others—appeared as a result of increased complexity of the objects of engineering activity or, to be more precise, as a result of increased complexity of designing such objects.

Tentatively, the middle of the 20th century can be accepted as the beginning of the non-classical stage in the development of technical sciences. It is sometimes said that, to characterise this stage, one must have a vision of the future of technical sciences. However, some essential features distinguishing technical knowledge of the classical type from non-classical technical knowledge can be identified already in these days. They differ in (a) the structure of theories; (b) the overall structure and organisation of non-classical scientific-technical knowledge, which is interdisciplinary in character and has a two-layer structure; (c) the mechanisms of emergence and formation.

The content of points (a) and (b) is discussed below. As for the institutionalisation of non-classical scientific-technical disciplines, its specificity consists in that, considered in the most general terms, they develop according to the schema "scientific movement—scientific discipline" (100, 40-43).

Methodological analysis of non-classical technical disciplines is rather difficult, as they are at a stage of theoretical evolution now, and the ideal of the organisation of their knowledge has not yet been fixed. The methodological situation in this case is similar to that which existed at the beginning of the 20th century, when the theoretical level of the classical technological sciences was not yet clear. But even the study of well-established aspects of some non-classical sciences shows that non-classical scientific-technological disciplines are a qualitative leap in the development of technical knowledge.

The problem of synthesis of complex technical sciences is one of the most important problems in the development, at the present stage, of both the sciences themselves and their methodological analysis. Synthetic processes in scientific knowledge, including technical knowledge, are closely linked with integrative ones. Integration of scientific cognition now attracts special attention of scientists concerned with the methodology of science, as reflected, e.g., in the materials of the Third All-Union Conference on the Philosophical Problems of Natural Science, where analysis of integrative processes figured quite prominently (26). The need has arisen in the methodology of science for strictly distinguishing between the phenomena and concepts of integration and synthesis of scientific knowledge. Without this, a study of the dynamics of modern natural science, in particular of synthesis and integration of sciences in non-classical scientific-technical disciplines, is practically impossible.

Integrative processes in science, even if we consider their epistemological aspect only, constitute a very complex structure. This complexity, the objective diversity of integration, is reflected in the great number of typologies of integrative processes built on different foundations. The essence of integration is unification of previously autonomous parts in a whole, in a system. In the sphere of knowledge, these parts are integral fragments of knowledge or scientific disciplines. It should be noted that integration does not necessarily result in an integral formation. In particular, a scientific movement as a form of the process of integration in science does not always culminate in the founding of a scientific discipline. A well-known example of a scientific movement which did not give rise to a scientific discipline is programmed instruction. A better known case of scientific integration (better known, perhaps, because of its success) is the cybernetic movement and the subsequent emergence of cybernetics. At present, we witness the extremely powerful systems movement which integrates different means and styles of thinking, different conceptions, values, etc.

A scientific movement develops into a complex science only if a synthesis is achieved of previously autonomous fragments of knowledge, methods and scientific disciplines. Synthesis is a further stage in the development of integrative processes in a scientific movement, characterised by the merging of autonomous and different forms of knowledge in a uniform whole.

As the final phase of a scientific movement, it is "a basis for the formation of a new, independent branch of knowledge" (26, 312).

Synthetic processes, considered in the epistemological aspect, are a further development and the result of integrative processes and, consequently, retain their most important traits. This is also true of the conditions of synthesis of knowledge in scientific disciplines. In this connection, the most adequate approach here seems to be the view that synthesis of knowledge in interdisciplinary sciences is determined in a dual manner—by the object of cognition and by activity-related factors (ibid., 170).

Among other complex (non-classical) technical sciences, the problem of synthesis of knowledge is analysed, e.g., in the methodology of ergonomic studies. Analysis of the practically generally accepted or, at any rate, the most authoritative viewpoint in this field shows that the solution of the question is, in brief, as follows. This solution is based on recognising the two-layer character of ergonomic knowledge, which consists of substantive knowledge and knowledge pertaining to research. The solution itself is formulated as follows: "Object knowledge brought in from other disciplines creates, through knowledge gained from research (always connected, in one way or another, with a certain practical task), a new organic whole" (100, 74; 101, 31). In other words, the addition that makes an agglomeration of knowledge in psychology, physiology, labour hygiene, etc., ergonomic knowledge proper, consists in the mode of organisation of the knowledge of ergonomics.

In this general formulation of the solution to the problem of synthesis of ergonomic knowledge, especially noteworthy is the link between researcher's knowledge and practical scientific-technological tasks. It is the task that specifies the schema of synthesising autonomous, different scientific disciplines into the new, systems quality of ergonomic knowledge. The task is, figuratively speaking, the crucible in which autonomous fragments of knowledge melt to form ergonomic knowledge proper. In view of our insufficient knowledge not only of the mechanisms of technological creativity but also of creativity in general, it is at present difficult to describe the process of transformation of integrated disciplines into the knowledge of ergonomics or to characterise the stages of this transformation. However, the nature of the task itself can be indicated, and important methodological and epistemological conclusions may be drawn. We have pointed out above the purposive nature of technological, practical tasks and problems. The goal is the core of practical and technological tasks. The goal may be said to be the organising principle in ergonomic knowledge, just as in systems engineering, systems design, and other non-classical technical sciences.

The structure of technical knowledge. Technical knowledge is a system of its pre-scientific and scientific forms, an ensemble of classical and non-classical sciences. Considered from the epistemological angle, the scientific-technological disciplines are in their turn integral structures comprising theories, hypotheses and laws, facts and ideas, technological tasks, etc. They can be either at the theoretical stage of development or at the empirical one, or else in between—in the process of theoretical formation.

As a type of human knowledge, technical knowledge is a schematisation of technical practice and of technics in the above sense. At the initial stages of development, in its initial structures, technical knowledge is not a schematisation (which is theoretical in nature) but rather an interiorisation—assimilation and transposition of external laws onto the internal, ideal plane of consciousness.

Being a dynamic, developing structure, technical knowledge produces, at a definite stage, technical sciences. The development of technical sciences results in the formation of specific integral structures within technical knowledge itself. The problem arises of characterising or describing this integral whole. What methodological approaches and epistemological characteristics are adequate in this case? Can this structure be, say, a "scientific region"?

Some authors believe that that is so, positing the theory of mechanisms and machines as the leading discipline of the region of technical sciences (114, 63). Since machines do not exhaust the whole of technics (in the sense of artificial systems), the general theory of machines cannot, in our view, be the key to the anatomy of knowledge in the sciences of the technical region. This approach ignores knowledge objectified in the artificial systems of modern technologies, such as the chemical or microbiological systems using microorganisms—various yeasts and bacteria—in obtaining oil, copper, uranium, etc. Chemical technology, to take an example, does without machines and even without implements of any kind. Bearing in mind the complex forms of integration of scientific knowledge that have been brought to light and analysed (92), we can perhaps best describe technical sciences as a complicative system.

As we have already mentioned, the most adequate unit of the methodological study of the system of technical sciences is a scientific-technical discipline. Emphasis on this unit in the methodological analysis of technical knowledge permits an effective description of various aspects of technical sciences, opening up, in particular, good prospects for the study of the structure of scientific-technical knowledge.

Methodological analysis of a scientific-technical discipline necessarily assumes the study of the formation of the theoretical schemata of that concrete discipline. In other words, one of the first questions that arises in this case is the question of the formation of theory in a concrete technical science. In technical knowledge, just as in scientific cognition in general, a theory is the principal structural unit (15, 110). At present, the central problem of epistemological, methodological research of scientific-technical knowledge is therefore the problem of technical theory—its formation, structure, functioning, etc. Analysis of the meaningful and formal aspects of a technological theory reveals deeper strata of technological knowledge than, say, the study of technological rules, which is the principal task of the epistemology of technological knowledge according to Bunge (172, 68).

Emphasis on technical theory as the central problem of the methodology of scientific and technical knowledge does not contradict the view, which we share, that the most adequate unit of methodological analysis of technical knowledge is a scientific-technical discipline. At present, there is a need for methodological studies in concrete scientific-technical disciplines, especially non-classical ones. The focus here must be on analysis of problems of theory in these sciences.

It is pointed out in the literature on the methodology of scientific cognition that, on the standard methodological approach, a theory is the basic structural unit of scientific knowledge. Although this approach is justifiable on the general methodological plane, it should be noted that as far technical sciences are concerned, even this standard model of scientificity has not been properly analysed so far. Considering the role of technical sciences in this age of the scientific and technological revolution, methodological research of theoretical schemata in the scientific-technical disciplines is necessary from the practical standpoint as well.

A promising approach to this issue is connected with the use of experiences in meaningful analysis of natural-scientific, above all physical, theories which have been accumulated in Soviet methodological studies. On the analogy of physical theories, the following components can be singled out in technological theories: theoretical (ontological) schemata, and conceptual and mathematical apparatuses. These elements naturally differ in content in technological and natural-scientific theories.

This approach to the study of technical theories is developed in a number of works (22; 111). Summing up the principal ideas of these studies and omitting some details and terminological differences that are irrelevant in the present context, we can say that three principal strata are singled out in technical theories: functional, "assembly-line", and structural. It should further be noted that, although a non-classical technical theory is built differently from a classical one, the principal components—ontological (theoretical) schemata, conceptual apparatus, and mathematical apparatus—figure in both cases. Unlike a natural-scientific theory, a technical theory does not contain new logical connections. The principal distinctive feature of a technological theory is the constructive function; such a theory does not just explain and predict; primarily, it ideally generates engineering objects.

In developed technical sciences of the classical type, the functioning of a technical theory creates the conditions for distinguishing between the engineering and the scientific form of activity in the scientific-technical disciplines. The constructive function of technical knowledge and of the sciences does not consist in their creating technical objects but in the production of knowledge ideally generating engineering objects. Technical theory creates schemata which are later implemented in machines, mechanisms etc. through engineering activity.

As a matter of fact, the structure and functioning of theories of the classical and non-classical types in the technical sciences can serve as models of the structure of the entire technical knowledge. The structure of scientific-technical knowledge is a system or complex of classical and non-classical technical theories. Technological, constructive-technological knowledge, and knowledge of practical methods constitute in this case the empirical basis of these theories (22, 45). The other concepts existing in the sphere of scientific-technological knowledge cluster round theories, acquiring a meaning only in the context and in terms of the language of technological theories.

Not all problems and mechanisms of the construction and functioning of technical theories have a clear solution, especially in non-classical sciences. That is especially true of the study of the "configuration" process in non-classical technical theories.

Let us consider some aspects of classical technical theory, where most problems have been solved and the mechanisms of functioning are more or less clear. In the process of the functioning of a technical theory engineering objects are ideally produced. The solution obtained is then transformed at the level of engineering activity, where the parameters that are secondary in terms of the ideal model (size, weight, and the modes of joining parts) are taken into account. It is at this stage, as a rule, that industrial designers join in the work on the object. This sphere of "second-order features" has been and often still is the field of artistic designing. In collaboration with the industrial designer, or without him, the product—the technical object—changes its form, its aesthetic properties.

This aspect is important from the standpoint of analysis of technical knowledge and its relation to technics, because changes of the object's form and aesthetic properties produce alterations in the engineering object through objectification of knowledge. In other words, there is a need for clarifying the specificity of knowledge thus objectified in technics and the place of this knowledge in the structure of technical knowledge in general. The artistic qualities of technics and the origin of these artistic qualities are the "subject-matter of special research into extremely complex problems that have so far been merely touched upon" (98, 24). The form-building situation is much the same; we are faced here with "numerous problems that are not yet analysed, far from clear, and some not even identified, in fact" (109, 1). We could cite a great many statements and striking remarks showing quite clearly that in working on the form of an engineering object, the industrial designer is guided by semi-intuitive considerations and empirical laws. The methodology of industrial design is at present at the very first stage of development; at best, we can speak here of an ensemble of practical methods used by designers.

This does not of course call in question cases of successful cooperation between engineers and designers, still less, the awareness of the need for such cooperation. Technical theories must ideally generate engineering objects with due attention to aesthetic and ergonomic requirements. In other words, it is necessary to construct theories in industrial design and ergonomics and to synthesise these theories with other technological theories in a general theory of technics. One of the first steps in this direction must apparently be reflection of aesthetic qualities of an engineering object in structural ontological schemata of technical theory.

In addition to what has been said here of the structure of technical knowledge, we can therefore describe it as a system of well-established classical and non-classical technical theories plus ontological schemata modelling the relationship between man and technics. These schemata, empirical in nature, belong to the sphere of industrial design and ergonomics. Although they reflect properties of an engineering object which are secondary in the eyes of the engineer, they still belong to the sphere of technical knowledge. At present, these properties are more and more taken into account in designing and constructing technical objects; the share of ergonomic and aesthetic knowledge in the overall body of technological knowledge is growing.

Analysis of the specificity of technical sciences raises the problem of the criteria of their scientificity. Just as in mathematics and natural science, the question naturally arises of the central criterion of scientificity in the technical sciences.

Scientific-technical disciplines of the classical type are obviously characterised by the features of intersubjectivity and systematicity. This is largely determined by the origin of classical technical sciences—their derivation from natural science, and by the high degree of the mathematisation of their theories.

Of these two features, intersubjectivity is the less graphically expressed feature of the scientific-technical disciplines of the non-classical type. This is largely due to the fact that theory in these disciplines is, as a rule, at the initial stage of development. Intersubjectivity, of which the basis is invariance of knowledge, reproducibility of its results, methods, etc., is very weak in industrial design; it is somewhat stronger in systems design.

The low degree of the intersubjectivity of industrial, "piece-work" design is largely determined by its nature, its closeness to art rather than science. Generally speaking, the grouping of industrial design together with non-classical scientific-technical disciplines is due to convention more than anything else, and to its kinship with systems design.

As for systems design, the intersubjectivity of its methods, results, etc., is fairly high. In recent years, however, the penetration of the principles and methods of artistic thinking in the sphere of systems design seems to have brought about a fall in the share of intersubjectivity in systems design. That does not signify a fall in the level of scientificity in this sphere of knowledge but rather its restructuring and the strengthening of links with the human sciences.

The intersubjectivity of other non-classical scientific-technical disciplines is higher; evidence of this is found, in particular, in the striking systematicity of these disciplines. The complex systems character of objects designed in these days determines the fact that systematicity of knowledge proves to be the most important methodological regulator of theory construction in these disciplines oriented towards a universal ontological schema. The latter is represented in the concrete non-classical theories by some version of the general systems theory, by the instruments and methods of the systems approach.

As has been pointed out above, science may be seen as well-formed if it effectively performs its principal function. For technical sciences—classical and non-classical—that principal function is the constructive one, which is concretely expressed in the existence of practical knowledge of the methods and theory of this science. The constructive function of technical theory constitutes its principal difference from natural-scientific theory, whose main tasks are explanation and prediction.

The criterion of truth—one of the principal criteria of scientificity—is of the greatest interest in the case of knowledge of practical methods. Fundamental results concerning the truth of ontological schemata of technical theories are accurate, to use a technical phrase, up to a sign in the study of the abstract objects of scientific-natural theories. The problem of the truth of the knowledge of practical methods forms the very core of the question of the truth of technological knowledge.

To what extent is knowledge of practical methods true, and what does it reflect? The criterion of truth in the scientific-technological disciplines is engineering activity, which acts as experiment in relation to technical knowledge. Does that mean that only those ideally reproduced engineering objects, various projects of systems design, ergonomics, systems engineering, etc., are true which were realised, the realisation including their engineering, designing, etc., finishing in reality? A positive answer to this question is fraught with the danger of the pragmatist interpretation of truth. The abovementioned ideal objects can be described as versions of relative truths. But what do these ideal essences reflect? It is hard to answer this question without distinguishing between purposive activity and its substantivised forms in the structure of technics. Those versions of practical recommendations on methods which are not realised, at least at present, reflect goals and goal-directed activity as an objective process.

The answer to the question of the truth of the knowledge of practical methods in technical theories and of their constructive function permits the formulation of the central criterion of scientificity in relation to technological knowledge. That criterion is constructiveness. Attempts have been made in methodological analyses of technical knowledge to clarify the specificity of technical knowledge through the study of their constructive-practical nature (see, e.g., (106; 114)).

One of the basic elements of our view of the constructiveness of technical sciences is the interpretation of the correlation between engineering activity and scientific knowledge in the sense outlined above. Constructiveness as the central criterion of the scientificity of classical and non-classical disciplines consists above all in conceptual reproduction of engineering objects in the functioning of technological theories. As far as the development of classical and non-classical technical theories is concerned, constructiveness is necessarily connected with effectiveness, which is perceived as optimisation and reduction to a minimum of the reproduction of ideal engineering objects that are not realised in technical practice.

3.4. HUMAN KNOWLEDGE

The principal component parts of human knowledge are the human sciences and artistic knowledge. The fact that artistic knowledge is part of human knowledge precludes the definition of human knowledge as a complex form of integration of science, although the human disciplines do show a tendency towards integration and even synthesis. Out of considerations of style and method, let us use the apt term "human-scientific knowledge" introduced in (19, 140), in the sense of the human sciences.

The unity of artistic and human-scientific knowledge is manifested, above all, in the fact that these forms of human knowledge study one and the same domain—man. Man, the world of man, the specifically human element—these are, as a rule, the definitions of the subject-matter of artistic and human-scientific knowledge. We are putting them on record here as they are quite satisfactory as the first step towards the clarification of the subject-matter of human knowledge.

The epistemological status of human knowledge. Human knowledge is one of the least studied problems in epistemology. It would be wrong to say that the reason for that is the fact that in his practical and cognitive activity man ignores the phenomenon of man himself. Nevertheless, there are obvious and fundamental differences concerning the subject-matter, methods of research, and forms of organisation of human knowledge; as a result, there are no generally accepted results in this sphere of epistemological analysis.

In concrete terms, this epistemological situation is expressed in the absence of a clear conception of the norms and ideals of the organisation of human knowledge. Existing philosophical-epistemological, methodological studies of human knowledge concentrate, in fact, on what human knowledge must *not* be.

Numerous cases, mostly interconnected, are cited in explanation of the existing situation. One of the most important of these is inadequate development of the theoretical level in the human disciplines and the weakness of methodological reflection in the sphere of human knowledge in general. This last point was touched on by Heinrich Rickert; as early as the beginning of this century he complained that scholars working in the "cultural sciences" (*Kulturwissenschaften*) "show little of the inclination towards methodological research which so richly rewarded the founders of modern natural science; or else when one finds deeper studies of the essence of their own activity..., they are isolated and limited to special fields" (180, 12).

The deep processes taking place in the core of modern science create real scientific premises for optimism, for progress in the study of human knowledge. The complex character of the problems solved at the time of the scientific and technological revolution increasingly involves linguistics, psychology, artistic knowledge and other areas of human knowledge in its applied aspects in the solution of tasks in industrial production. These tasks stimulate the development of complex (non-classical) technical sciences in which mathematical, natural-scientific, social and human knowledge is synthesised. As a result, the acute problem inevitably arises of the specificity of human knowledge both in the scientific and the artistic form.

In connection with the question analysed here, it should be noted that, in general, an important aspect of the dynamics of knowledge, of its restructuring, is the strengthening of reflection in terms of human knowledge, the strengthening of the influence of human thinking on natural science, among other things. As G. Gachev aptly put it, "a possible contribution of human sciences to the development of natural ones" is now on the agenda (63, 109). The effectiveness of this contribution will be much greater than the one we witness now if the processes of interaction between human and other forms of knowledge are subjected to a deeper methodological, epistemological analysis. Well-known facts concerning the substantial influence of aesthetic, ethical and, generally speaking, human elements on the creative work of outstanding scientists should become the object of intense methodological studies.

What we have in mind is precisely methodological, epistemological analysis of the influence of human knowledge on natural science and mathematics. Without calling in question the depth and value of the available experiences of studying the impact of aesthetic and ethical factors on scientific cognition, we must note that this problem is studied, as a rule, in psychological, aesthetic, or philosophical-culturological terms. Realising the difficulty of distinguishing between the processes of cognition and creativity, we must stress the need for studying the interconnections between the natural-scientific, mathematical and human knowledge in its logical-epistemological, methodological aspects. To paint a life-size epistemological portrait of science, methodological studies must place greater emphasis on the epistemological structures, forms and methods, the scientific and artistic pictures of the world, scientific theory and artistic image, modelling in natural science and human knowledge, analysis of abstract objects, the function of explanation in science and art, etc. What we need is not only psychological and other analyses of creativity but also epistemological inquiry into the forms of knowledge, above all human knowledge, analysis of conceptual borrowings from human knowledge by natural and technological sciences.

The changed situation in the methodology of sciences should also be seen as one of the factors favourable to effective analysis of human knowledge. It became obvious in recent years that a model of science in general could not be built on the basis of mathematics and natural science only. Thus the human sciences have acquired the right to exist as objects of the methodology of scientific knowledge, and it is generally acknowledged that the hammer of scientificity extracted from physical-mathematical natural science no longer threatens to crush them.

The causes of the changes in the status and functions of human knowledge, of its reappraisal in the methodology of scientific knowledge, are rooted in the social structure, in the needs of social development. Methodological problems are emphasised in science during critical periods of collapse of old concepts and methods and generally during periods in which science attains new levels of development. In natural science and mathematics the links between critical situations and social cataclysms are less noticeable, they are practically not realised by scientists working on methodological questions of the concrete sciences going through a crisis. But that does not mean that these correlations are non-existent.

Regardless of any problems of internalism and externalism, let us note that in modern science critical states of society are reflected not only in the moral appraisal of the results of scientific studies but also in the theoretical constructs proper of natural science and mathematics. Owing to the substantive orientation of cognition in physical-mathematical natural science, the mechanisms of this influence often disappear from the field of vision of the scientists themselves, from their methodological deliberations over the problems of their science. The theme of crisis may be said to penetrate the surface of natural science and mathematics in a mediated manner, through psychological motives of creativity, through the scientist's philosophical and worldview principles.

It is from this angle that the links, say, between the ideas of Dostoyevsky and Einstein, of Kierkegaard and Bohr are studied. With certain reservations of the tentative nature of this surmise concerning links between the ideas of outstanding thinkers in human and natural-scientific fields, it is noted that "these links consists in the fundamentally non-observable mechanism of psychological resonance rather than in the borrowing of *concepts* (italics mine.—*A.K.*)" (47, 600).[9]

Crisis states of the human individual, of personality, determined by the crisis of social relations are the motifs of the work of both Dostoyevsky and Kierkegaard. It was precisely the ideas of disharmony in the relations between the individual and the social whole in the work of Kierkegaard and Dostoyevsky that found a "psychological resonance" in the physical theories of Einstein and Bohr and, moreover, became part of their inner logic (ibid., 589). Thus crisis states of society are reflected in natural-scientific theories in a mediated manner, through philosophical meditation on the individual.

In human knowledge, in the human sciences in particular, the links with social crises are obvious and, moreover, just as obvious is the dependence of crisis states of human knowledge in its principal forms on social causes. The reason for this lies in the specificity of human knowledge, its substantive orientation towards man. The endless chain of the crises of capitalist society, which began with its entry, in the last third of the 19th century, in the epoch of imperialism, determined the kaleidoscopic succession of schools, trends and systems in bourgeois art and in the methodology of human sciences.

The famous Spanish sociologist and art critic José Ortega y Gasset was one of the first to have reacted to the new situation in the human sphere. "The system of values," he wrote in his

The Modern Theme, "by which its activity was regulated thirty years ago has lost its convincing character, its attractive force and its imperative vigour. The man of the West is undergoing a process of radical disorientation because he no longer knows by what stars he is to guide his life" (177, 79). Of all the forms of cognition, the crisis state of society is first of all reflected in the artistic form, in the images of art. Now, what are the generic, the most characteristic features of this new art? The answer to this is contained in the title of Ortega y Gasset's other well-known work, written in 1925—*The Dehumanization of Art*, where he wrote this, among other things: "Wherever we look, we see the same thing: flight from the human person" (176, 30).

One aspect in particular can be singled out in the complex interconnections of the cultural, spiritual crisis of bourgeois society and human disciplines. It is associated with the fairly popular and regularly revived idea of the overcoming of nihilism and decline of culture through reorganising and restructuring human sciences. We find different versions of this idea in Edmund Husserl (147), Bronislaw Malinowski (166), in French structuralism, Victor Frankl's logotherapy, etc. This approach to the crisis phenomena in culture appears more acceptable than, say, the Stoical position of Max Weber, who proposed that we accept the "epoch of crisis" as our fate and historical necessity. This sort of methodological activity, however, will never provide a spiritual and material foundation for human life.

In Marxist epistemology, methodological reflection on human knowledge unfolds in a basically different context—the context of socialist humanism. Despite the external similarities of form (the restructuring of human knowledge in the process of its development, a more profound study of man in this connection, etc.), the concern for the methodological questions of human knowledge in Marxist epistemology is determined by the tasks of bringing up the new man, and not by social crises.

Yet, elements of the crisis situation in some spheres of the society, for instance in the USSR in the early 1980s, may quite seriously stimulate an interest to man (the "human factor"), the cultural sphere in general and thus to human knowledge and its methodological problems.

Epistemological explication of human knowledge is always oriented towards a certain epistemological ideal or standard of science. The broad (weak) version of the concept of science formulated in A. Rakitov's work (81, 128-130), appears to hold promise for the methodological analysis of the specificity of human knowledge in its scientific and artistic forms.[10] The core of this version is a set of features specifying a certain model of science which is not rigidly oriented towards the mathematical and physical standards of scientificity and at the same time is different from everyday knowledge.

Orientation towards a certain epistemological ideal of science, of course, opens up some prospects with regard to the investigation of human, and particularly human-scientific, knowledge. A fuller clarification of the status of human knowledge in all contexts can, however, only be obtained by studying knowledge as a whole. The interpretation of cognition as such forms the basis for working out the specificity of human-scientific and artistic cognition.

The Marxist theory of knowledge contains a number of definitions of knowledge which lay claim to general validity and executability for all spheres of cognitive activity. One authoritative publication states, "Knowledge is the subjective image of the objective world" (45, 125). What is most conspicuous about this interpretation of cognition is its connection with Lenin's well-known proposition, "Sensation is a subjective image of the objective world" (51, 14, 119).

Can this proposition of Lenin's, correct as it is for sensation as an elementary form of psychic reflection and knowledge, be given the status of a general definition of knowledge? Quite apart from the familiar difficulties with the explication of the "world", starting with neo-Kantianism and Dilthey, the question arises: what is to be done in the final analysis with the subjective world, the subjective image—consciousness? Can they (the question is purely rhetorical since they obviously can and must) be the object of cognition? But what in this case constitutes the basic formulation of cognition? Is knowledge the subjective image of the subjective world? It is quite evident that this approach embodies manifest and non-manifest generalisations of the natural and technical sciences of the classical mould and that it cannot serve as a basis for the explication of human cognition.

In dealing with the theory of cognition, we are faced with a situation where it is in principle impossible to give an exhaustive definition of knowledge once and for all. The point is not that any definition is always limited. To provide a once and for all definition of knowledge which lays claim to exhaustiveness means in essence to limit the process of cognition to the level existing today, of which the proposed definition will be a kind of copy.

Lenin pointed out, "a full 'definition' of an object must include the whole of human experience..." (51, 32, 94). This is all the more correct when applied to an object like cognition.

How can experience be introduced into a full definition of cognition? The obvious way is to correlate cognition and experience. Soviet studies contain both interesting methodological substantiations of this approach and convincing examples of its implementation, its concretisation applied to material from the realm of physics and the technical sciences (22, 111).

Works by the authors referred to present cognition, here in the context of physics and the technical disciplines, in the form of the schematisation of the object aspect of experience. Experience is treated as an "intricate web of different acts of transforming objects, where the products of one activity become the starting components for another" (66, 138).

By developing this approach, cognition as a whole, in its scientific and, indeed, extra-scientific forms and aspects, may be treated as a schematisation of experience. Since knowledge is entirely dependent on experience, the development of the latter determines the development of the conception of knowledge. By linking experience and cognition and by introducing experience into the definition of knowledge in this way, the limitations on concepts of knowledge are constantly removed and the notion of knowledge is developed. The key to an explication of human knowledge and the prospects for defining its status are to be found by observing and clarifying its connection with the subjective aspects of experience.

Cognition as the schematisation of experience is a process of creating abstract objects. In the case of scientific cognition, these are ideal objects; with artistic cognition, they are artistic images and their elements, whilst in the case of commonplace cognition and myths, they are their abstractions, etc.[11]

The subject-matter of human knowledge. In keeping with the broad version of the concept of science accepted here, the first problem of epistemological explication of human knowledge is that of its subject-matter. At the present stage of the methodological analysis of human knowledge, the question of its subject-matter is largely a question of the possibility of the existence of human knowledge. For the human sciences and artistic knowledge to exist as specific forms of human knowledge, objects must exist that are involved in the sphere of man's practical activity and are not studied in other forms of knowledge.

In other words, the extensive sphere of man and the world

of man must have aspects that are studied by human knowledge alone. If that is not the case, the obvious fact of activity in human cognition will be perceived as a scholastic exercise or game rather than as schematisation of practice. This last choice, that of game, is unacceptable even for artistic knowledge, to say nothing of the human sciences.

The view that the subject-matter of human knowledge is man in the entire diversity of his practical activity must therefore be concretised. Above all, that is necessary for specifying the subject-matter of the human and the social sciences, for in this interpretation the boundaries between them are vague, which gives rise to various versions of identification of the human and the social disciplines or to the incorporation of the social sciences in the human ones. Epistemological explication of the subject-matter of the human sciences also creates the premises for differentiation between knowledge in these sciences and artistic knowledge.

The identification of the social and the human sciences is mostly expressed in the fact that, although differences between human and social sciences are postulated or implied, concrete lists of the human disciplines are drawn up which also comprise the social sciences, at least the most advanced of them. For instance, such works as (24, 6; 191, 11) include among the human sciences history, sociology, psychology, anthropology, archaeology, economics, the philological sciences, politics, ethnography, the ensemble of the sciences of art, parts of philosophy, etc. The obvious typological similarities of these viewpoints and the just as obvious absence of mutual influences between them indicate the wide currency of this approach to the problems of the human and the social sciences.

All sciences are human in the sense that they "close on" man, being engendered in the process of man's practical activity, their results having value only insofar as they are used, sooner or later, in man's practice. In this sense, the natural, social, technical, and the human sciences proper are all sciences of man. It is this deepest essence of science, determining its general tendency of development, that will bring about a situation in which "natural science will in time incorporate into itself the science of man, just as the science of man will incorporate into itself natural science: there will be *one* science" (59b, 3, 304). But a unified science whose only and universal subject-matter is man is a thing of the future. This future should not be brought closer to the present by ignoring (mostly unconsciously) the actual situation in present-day science. Analysis of the existing forms of scientific knowledge reveals, in the first place, difference between the subject-matter of concrete sciences.

What is the subject-matter of the human sciences? "The human sciences," wrote Mikhail Bakhtin, "are sciences of man and the specificity of man" (7, 285). It is not man in the entire diversity of his manifestations, but the human in man that is the main feature of the subject-matter not only of the human sciences but also of artistic knowledge. To solve the problem of the human in man, and thus to specify the subject-matter of human knowledge, would mean to answer, to a great extent, the principal questions which philosophy and science have faced since the moment of their emergence. We are not setting ourselves the task of investigating the specificity of man in its entirety—we shall merely comment on some essential aspects of this specificity, some tendencies and stable links in its functioning and development important for analysing the subject-matter of the human sciences and artistic cognition.

The specificity of the human must apparently be sought for in the content of the social qualities and characteristics of man. Certain anatomical traits can be identified that are characteristic of man alone, but that is hardly sufficient ground for regarding them as the specifically human in man. Neither can these features, inherent in all members of *Homo sapiens*, be described as the subject-matter of human knowledge, for they are not studied either by the human or by the social sciences. It should further be noted that these biological characteristics of man follow from his social qualities.

How can the specifically human be identified in such a complex system as man? In our view, it would be promising to use in this case the heuristic aspects contained in the methodology of materialist dialectics. In particular, bearing in mind the coincidence of the historical and the logical, one might trace the evolution of the correlations between the human element and human knowledge and thus attain theoretical-conceptual plane of the problem. Engels wrote this on the coincidence of the historical and the logical: "The point where this history begins must also be the starting point of the train of thought" (69, 225). If the problem has not crystallised in the logical-theoretical aspect, and it is not clear where the train of thought must begin, one should turn to the history of the question, where the stable, recurrent connections are the analogue of the logical consideration of the problem.

Since antiquity, that is, since the time of the emergence of science and philosophy in its classical form, knowledge of man has been accumulated not only in philosophy, theology, etc., but also in the liberal arts (*artes liberales*) and in the *studia humanitatis*. It is with the *artes liberales* and *studia humanitatis* that the formation of the specifically human, of humanity, and later their cognition, began.

Unlike the auxiliary arts (*artes vulgares*), the liberal arts did not require physical effort. This old classification of the arts, very old, very popular, and very tenacious, was an expression of the aristocratic social system of antiquity with its disgust for physical labour, the lot of slaves. In the culture of classical Greece, the link between the liberal arts including elements of human knowledge with the specifically human is purely external: the liberal arts were the sphere of activity of the free man, that is to say, of man as such, in the context of classical culture, for the slave was not human.

In late antiquity, mostly in Cicero, the liberal arts were linked with humanity (*humanitas*), which was taken to mean high spiritual culture, a higher type of education accessible only to the aristocratic upper stratum of Roman society. That type of education and culture was attained through the study of the free arts. Aulus Gellius' comments clearly point to the close links between the liberal arts and man: "The pursuit of that kind of knowledge, and the training given by it, have been granted to man alone of all the animals, and for that reason it is termed *humanitas*, or 'humanity' " (140, 456-457).

The transformations which had occurred in the liberal arts by late antiquity are of fundamental character and highly important for the birth of human knowledge. *Humanitas* as knowledge grew out of the (free) man's study of the liberal arts. This study yielded knowledge as its result. In late antiquity this knowledge came to be *identified* with man, with the human element. Knowledge resulting from the study of the liberal arts became knowledge of humanity, not just a result of the free man's study of the liberal arts. It became knowledge of the human, for its subject was man, that is, free man, not a slave. The carrier of the knowledge of the liberal arts became its object.

Among late Stoics and early Christians, a different conception of *humanitas* evolved, one that was close to Greek "philanthropy" and had a bearing on any man regardless of origin or education rather than certain elitist strata of society. Beginning with early patristics and up to the beginning of the Renaissance, Christian theology developed the conception of *humanitas* as a quality inherent in all men, as a particle of the divine in the soul of man. The principal instrument of attaining *humanitas* in the sense of a particle of the divine became faith and the study of sacred texts.

The liberal arts receded into the background in the study of humanity; moreover, as an element of education, they were reduced to the role of an auxiliary instrument in the exegesis of Biblical texts, a preliminary to the study of philosophy. Still, human knowledge represented in the liberal arts was not arrested in its development. Martianus Capella (the first half of the 5th century) "continued the work of classifying the sciences and established the so-called system of the seven liberal arts (*septem artes liberales*), later improved by Boethius and Cassiodorus" (21, 56).

Developing the ideas of Martianus Capella, Boethius divided the seven liberal arts into the quadrivium (arithmetic, geometry, astronomy and music) and the trivium (grammar, rhetoric or literature, dialectics or logic). Boethius was the true founder of the Latin mediaeval quadrivium,[12] but he also did a great deal for the development of the trivium, especially for logic. Concentrating his efforts on the disciplines of the quadrivium and uniting the remaining liberal arts in the trivium, Boethius sowed the seeds whose fruit proved to be more valuable than the results he obtained in the field of the quadrivium. Whether he realised it or not, the trivium concentrated and organised as a set of special disciplines that human knowledge which was destined to play an outstanding role in the Renaissance.

The epoch of the Renaissance was characterised by the emergence of the new personality. A new conception of man and humanity was forged in the ethics and philosophy of the humanists in the struggle against the mediaeval interpretation. The thinkers of the Renaissance stressed in man his individual self-consciousness, emotional, moral, and personal aspects.

The new conception of man and humanity (in the sense of *humanitas*) entailed a broadening of the liberal arts. These arts, especially the trivium, were transformed into *studia humanitatis*, and the scope of the latter was gradually extended. Various fragments of knowledge substantively correlated with features recognised by the humanists to be human—dignity, feelings, etc.—were the material out of which an essentially new corpus of the *studia humanitatis* was formed. In the first place, ethics, poetry, etc., were included here. Thus renovated in content, the *studia humanitatis* became an instrument of educating the new man, a spiritually free personality.

In the epoch of the Renaissance, the range of the traits of *humanitas* (humanity) was defined, and at the same time the integral structure of the *studia humanitatis* took shape—an ensemble of knowledge substantively oriented towards *humanitas.* It must be said, though, that from the very inception of the *studia humanitatis* the trend became apparent towards differentiation within *humanitas,* towards distinguishing in it various interconnected yet heterogeneous phenomena.

The development in the 17th and 18th centuries of physico-mathematical natural science made a decisive, and still persisting, impact on the cognition of the phenomenon of humanity. Characteristic of the formation of classical natural science was liberation from all the anthropomorphic elements, including elimination from the subject-matter of the individual-personality elements which obscured the pure image of nature. Important conditions of the possibility of the scientific revolution in the 17th century were the destruction of the classical and mediaeval concept of the cosmos and geometrisation of space, i.e., the replacement of the concrete space of pre-Galilean physics by the isotropic and homogeneous space of Euclidian geometry, in which the terrestrial was indistinguishable from the celestial, etc. In the scientific picture of the world and in the worldview oriented towards classical science, the individual along with his self-consciousness was not only driven from the centre of the world where he had placed himself during the Renaissance but literally could not find a place for himself in the world as it was seen at the time from the scientific standpoint.

The views of the first humanist Francesco Petrarch or of Blaise Pascal can illustrate the transformation in the self-consciousness of the individual in connection with the formation of classical science. The vertical destruction of the mediaeval cosmos placed the man of the Renaissance in the centre of the world. Man, who actually took the place of God, became the principal object of study for the humanists of the Renaissance. "What is the use—I beseech you—of knowing the nature of quadrupeds, fowls, fishes, and serpents and not knowing or even neglecting man's nature, the purpose for which we are born, and whence and whereto we travel?" asked Petrarch (190, 58-59).

The geometrisation of space and the new conception of time closely connected with it led to the final collapse of the cosmos as man's native house where he dwells without fear. The place and destination of man in the world were reinterpreted in the scientific worldview oriented towards mathematical natural science in the 17th and 18th centuries. The humanist attitude of the times was clearly expressed by Pascal, especially in *Thoughts on Religion*.

Humanity, forced out of the field of scientific cognition (of classical science) in the 17th, 18th, and early 19th centuries, was intensively studied in art, i.e., in artistic cognition. The art of that period also underwent significant changes. By 1750, the conception of art as creation of beauty rather than creative activity subject to definite rules finally asserted itself. Charles Batteux set apart the so-called fine arts: painting, sculpture, architecture, music, poetry and dancing. Previously, at the end of the 17th century, logic and arithmetic were transferred from the arts to the sciences.

In the 17th-19th centuries the tendency gradually increased towards distinguishing two aspects of *humanitas*: on the one hand, the anthropological, generic features of man, and on the other, the emotional-moral qualities of man concretised in terms of personality and individuality (13, 145-155).

By the mid-19th century, an independent and stable sphere of humanity became fairly clearly defined; it covered, in the first place, such phenomena as personality, individuality, the human "I". In philosophy and the social and biological sciences, which had more or less taken shape by that time, the elements of this sphere (as, e.g., the individual) were interpreted in terms of man's generic characteristics.

Obviously, the specificity of this domain is not fully exhausted in this way. The study of the same field by artistic consciousness is clearly inadequate. The need arises for a *scientific* study of the problems of personality, individuality, the human "I", etc. In the mid-19th century, this was most acutely realised by Kierkegaard and Feuerbach, who almost simultaneously attacked Hegel embodying at that time the philosophical-scientific approach to problems of personality and individuality. However, neither Kierkegaard nor Feuerbach succeeded in substantiating an integral theory of the specifically human, of humanity, and in creating in this way the foundation of a philosophical science capable of adequately handling the problems of *real* human beings.

The unitary field which took final shape by the middle of the 19th century and was concretised in such concepts as personality, the human individuality, "I"[13] was connected in the sphere of scientific cognition with an ensemble of sciences including philosophy, the social sciences, and humanities—all those disciplines in which knowledge is substantively directed towards man. Humanities was a very amorphous structure, in which not only philosophy and a number of social sciences were included but also art.

An excursus in the history of human knowledge and the development of knowledge of man in general reveals that in antiquity, in the Middle Ages, and in the Modern Times, the specifically human elements and humanity were understood in quite different ways. Depending on the concrete interpretation of this phenomenon, accents are placed on different disciplines, and the content of human knowledge varies. Now, what disciplines or fragments of disciplines included in the ensemble of the social sciences, philosophical sciences, and humanities, can be integrated, along with artistic knowledge, in human knowledge?

The solution of this question is made difficult by the fact that personality and individuality are social phenomena; they are social characteristics of man. Moreover, society is a specifically human structure. Man himself is nothing but an "ensemble of the social relations" (59b, 5, 4). In reality, which is independent of knowledge, personality, etc., as the "specifically human" element, forms an indissoluble, interpenetrating unity with socium, with culture. For this reason, in particular, the sociologist "taking the definite social relations of people as the object of his inquiry, by that very fact also studies the real *individuals* from whose actions these relations are formed" (51, 1, 406).

In consequence of this, each of the social and philosophical sciences includes knowledge whose content is the ensemble of problems connected with the human individuality and personality. The indissoluble unity and interpenetration of the "specifically human" and the social condition the unity of human and social knowledge. Human and social sciences are reminiscent at present of the two-faced Janus: we are dealing with either human or social knowledge in these disciplines depending on the aspect of man and his social characteristics considered.

The human sciences (human-scientific knowledge) are at present in their formative phase. The process of formation includes the setting apart in some well-established disciplines (linguistics, history, psychology, sociology, etc.) of such branches or disciplines as the psychology of personality, the psychology of emotions, the psychology of individual differences (differential psychology), the sociology of personality, linguostylistics, text linguistics, ethnoculturology, etc. The present state of the human sciences or, to be more precise, of the knowledge contained in these sciences, is a process of differentiation from social cognition—a process that is, so to speak, internal in character: the human sciences are derived from the social ones. The process is in a sense the reverse of what took place early in the 20th century, when the social disciplines separated from human knowledge (humanities).

Human and social knowledge form a unity. The duality of the content of knowledge in this unity, its indefiniteness and mutual transitions are registered in epistemological studies in the concepts "social-human knowledge" or the "social-human sciences" (20; 72; and others). The tendency of the development of this, figuratively speaking, "centaur" consists in that such sciences as psychology, linguistics, history, sociology, and others more and more turn to man. This is manifested in the emergence of such disciplines, derivative from the above, as the sociology and psychology of personality, in the development of methods of analysis of individual cultural phenomena—application, e.g., of mathematical methods in the studies of individual characteristics of various texts—philosophical, artistic, etc. These processes apparently express the concretisation of knowledge in social disciplines.

In the analysis of differences between human and social sciences it is important to realise something that we have already formulated above: all sciences are human in the sense that they are rooted in and applied to human practice. More human than others are such sciences as linguistics, psychology, or culturology—sciences that study phenomena inherent in man alone: language, consciousness, culture. And yet the expression "human knowledge" should be retained as a designation of artistic knowledge, too, as well as those emergent scientific disciplines which are derived from psychology, the social sciences and philosophy, and whose content is the study of problems of the human individuality and personality.

The reason for singling out human knowledge as an independent area does not consist in the fact that there is a powerful tradition of distinguishing between social science and humanities, especially in the English-speaking countries, or in the fact that methodological studies of scientific cognition often assert the fundamental difference between the human and the social sciences. The concept of humanities has not yet become crystallised—it is marked by a complete absence of a consensus in the definitions of the human sciences and in the views of their correlation with the social sciences and artistic knowledge.

Frequent assertions of a fundamental difference between human and social sciences are not substantiated by theoretical-conceptual arguments.

The area of the human is extremely extensive and dynamic; its expansions and contractions determine the dynamics of the social and human disciplines. Referring to the mobility of the limits of the human, Engels wrote of "a difference in the degree of ... humanity" (59c, 118).

In the extensive sphere of the human, the domain of the specifically human is singled out—a domain that is not fully covered by the theories and methods of the social sciences, psychology and philosophy. Mikhail Bakhtin repeatedly expressed the idea that personality and thing are the limits of cognition (7, 363, 367). The subject-matter of all sciences lies between these two limits. The formation of the human sciences, that is to say, their separation from philosophy, psychology and the social sciences, expresses, among other things, a desire for the study of the individual as one of the limits of cognition. The specifically human, which is concretely expressed in personality and human individuality, is one of the two points attracting knowledge that is scattered by the centrifugal force of integration in philosophy, psychology, and the social sciences. The second point or limit of the development of these sciences is the "thing"—impersonal, subjectless phenomena.

In the study of the human, the accent in philosophy, psychology and the social sciences, with the exception of branches integrated under human knowledge, is on generally valid, objectified, superindividual and impersonal aspects of reality. The human sciences, human knowledge in general, strive towards cognition of the individual and personal in all their fullness and concreteness. They draw their knowledge of that from the established disciplines, in the first place from those listed above, integrating them in an independent area of human knowledge. Thus all sciences are human to the extent that they comprise under their subject-matter the specifically human, above all the human individuality and personality in all their fullness, including their separate, unique manifestations.

Taking into account the duality (in the above sense) of knowledge, and primarily the number of purely human sciences that have not yet become separated from it, the share of human knowledge proper, we shall bring together linguistics, psychology and culturology under the heading of the human sciences. Human knowledge proper is still essentially interconnected with other fragments of knowledge in these sciences. Human disciplines as such have not yet separated from them to the extent that their applied aspects or, problems of the theoretical and practical level in them could be discussed outside their connections with the base sciences, the more so that in these base sciences themselves—linguistics, psychology, culturology, etc.—problems of the theoretical level and applied aspects are highly debatable.

The domain which acts as a kind of integrator of human knowledge is a multilevel system. Its complex hierarchical character is conditioned, in the first place, by the complex and polystructural nature of the basic components—the human individuality, "I", and the personality. For instance, in psychology alone four principal substructures of the personality can be specified: (a) orientation; (b) experience; (c) psychical processes; d) biopsychical properties (68, 196).[14]

The human individuality is also a complex polystructural entity characterised by integrality, separateness, uniqueness, consciousness, and creative potential (82; 87). Despite the absence of rigorous theories, abundance of methodologically weak or purely descriptive conceptions, the human "I" has also been shown to consist of autonomous substructures: self-identity, the ego, and the image of "I" (42, 31).[15]

The human individuality, being an active principle connecting all the social relations, reproduces these relations, manifesting itself in this way—among other things, in texts. The text—not only as a written record of speech but, in a broader sense, as a coherent sign complex, as a meaningful, sense-carrying system—is the next level of the subject-matter of human knowledge, derivative from personality and human individuality. In this interpretation, texts are also the subject-matter of the human sciences (7, 281; 181, 13).

The structuredness and stability of the integral systems of human individuality and personality, and of their manifestations in texts are nothing but forms of repetitiveness in the domain of human knowledge. They are also ontological foundations of intersubjectivity and reproducibility of the results in human knowledge. The system-structural organisation of the domain of human knowledge, characterised by stability and repetition, determines the possibility, given the truth of knowledge, of its cognition in the human sciences, as the criteria of scientificity in the sense of Chapter 1 will be satisfied.

The dialogue quality and truth of human knowledge. The specificity of the domain of human knowledge determines the specific features of its principal types—the human sciences and artistic cognition. A major characteristic of human knowledge is dialogue which, like the sense of expressions, is inseparably linked with understanding as a function of human research.

Dialogue has been a familiar feature in European science and philosophy since the days of Socrates and Plato, at the least. Dialogue in the sense specific for human knowledge first emerges in well-developed form in the humanist thinking of the Italian Renaissance which grappled with the problem of creative synthesis of different cultural heritages—those of antiquity, Christianity, its own, and so on. L. Batkin characterises humanist dialogue as "conflict of different minds, truths, dissimilar cultural traditions constituting a unitary mind, a unitary truth and a common culture" (6, 137). The humanist dialogue inherently interprets synthesis as retaining the truths of two arguing sides of the unitary truth. In other words, for the humanists of the Italian Renaissance, the unitary truth emerging as a result of synthesis in a dialogue is a plurality of truths. In the humanist dialogue, as Batkin pointed out, "*there is no development*—there is invariance of truths" (ibid, 168). There is thus no development in it towards the absolute or, to be more precise, the only truth; instead, there is recognition of the right to one's own interpretation of the objective truth, emphasis on relative truth.

This view of synthesis in dialogue was determined by the nature of humanist culture with its profound realisation of individuality—the individuality, first and foremost, of cultures and the human personality, and recognition of their rights to a view of their own and ultimately of their right to exist.

Differentiation of knowledge of art and of science in the Modern Times—a process that also involved knowledge in the human sciences—determines the need to consider dialogue in human knowledge with reference to its two basic forms, the human sciences and artistic cognition. Despite certain common elements, dialogue in the human sciences and in artistic cognition obviously manifests certain differences in the tendencies of development. The common feature of dialogue in both cases is its indissoluble link with understanding as a function of inquiry and with the problem of truth. The tendencies towards differentiation in dialogue are also connected with the relation to truth.

In artistic cognition and in some human sciences, dialogue as a form is a consequence of recognition of equal rights to the truth for both sides interacting in the dialogue. Dialogue results from the meeting of two subjects (consciousnesses), one of which, the object of cognition for the other, cannot remain a mute thing. In M. Bakhtin's terminology, "voice" is necessary to the consciousness which appears in dialogical cognition as the object, in order to impart to the cognising subject, through dialogue, the meaning that is indissociably linked with the uniqueness of consciousness as an object of cognition. Without the "voice", the individuality and uniqueness of the subject as the object of cognition is inaccessible either to scientific cognition with its general concepts, schemata, etc., or to understanding with its orientation towards meaning.

Dialogue as the only form of cognising the unique, the one and only subject, implies listening to the subject's "voice". A mònologue rejects the equal rights of consciousnesses in relation to the truth (7, 309). Dialogue assumes not only equal rights to the truth but also, in fact, recognition as truth of everything announced by the "voice" of individuality. In its exaggerated form, this view is expressed in the concepts of truth as communication in Karl Jaspers or truth as intersubjectivity in Gabriel Marcel, leading to the pluralism of truths. The slightest exaggeration of the rights of the sides to truth leads to dialogue becoming, in Kierkegaard's words, a sum of "truths for me". In other words (Kierkegaard's again), "the individual is in the right even when he stands in this relation to untruth" (152, 190).

Dialogue in this interpretation is based on two elements: (a) conception of the subject—the object of cognition—as a unique individuality; (b) impossibility of adequate reflection of this individuality in terms of scientific (actually, natural-scientific) knowledge. Dialogue in this sense occurs, within limits, say, in culturology, where the uniqueness of cultures—the one that is cognised and the one that provides the context for the cognition of the former—is the basis of dialogue in culturological knowledge.

Human sciences like linguistics or ethics characteristically tend to interpret dialogue in a sense that is closer to natural science and mathematics. As we see it, it is this sort of dialogue that more and more asserts itself in the developing human disciplines. The fundamental difference between dialogue in this sense and its first interpretation lies in the nature of the synthesis of the two viewpoints interacting in the dialogue. In the first case, the dialogical relations left the truths of the two sides in polemics intact, whereas the second version of dialogue implies their changes, corrections, and even elimination (in case of falsity) in the synthetic movement to an objective, generally valid truth.

Dialogue and polyphony of human knowledge in the second sense of the term more and more tend towards transformation into two-hypothesis and multi-hypothesis discussions respectively (55, 56-57). Discussion is here interpreted in a logico-epistemological sense rather than as a form of communication or exchange of opinions in the communicative sense. On the logico-epistemological plane, discussion is a form of creative scientific cognition, of movement of knowledge to truth through conflict and mutual enrichment of theories, conceptions, ideas and viewpoints.

Considering these processes of divergence in the dialogues of human knowledge, we should perhaps think in terms not only of dialogue but also of discussion and, better still, probably of the multiformity of human knowledge. The latter is more adequate than dialogue in the case of the human sciences.

Analysis of the problems of dialogue in human knowledge shows that "the human sciences, the sciences of man and society, require the development of criteria of truth and objectiveness that are not identical with or reducible to the criteria of natural-scientific knowledge" (2, 233). The weak versions of the general validity of truth, as illustrated by the conception of truth as communication in Jaspers, stem from the view of personality as the subject-matter of human knowledge. Generation of truth in the act of communicating existences, and plurality of truths do not simply result from the uniqueness of personality—they follow from the uniqueness of an absolutely passive, static, structureless personality that does not manifest itself in anything but the communicative act.

True cognition of a unique culture or human individuality can be generally valid if we do not postulate in advance their absolutely static and structureless character. The stability of the structures of the manifestation of personality or human individuality, and their own systemic character, are the basis of the general validity of the truths of human knowledge.

Applied aspects of human knowledge. The fundamental and applied levels or aspects can be singled out in both types of human knowledge—in the human sciences and in artistic knowledge. The study of the fundamental and applied characteristics of human knowledge reveals the same problems and solutions as figure in the other forms of knowledge—mathematics, natural science, etc.

It was mentioned in the section on technical knowledge that

knowledge acquires applied features only when it is applied to the solution of production tasks. Accordingly, such a science as psycholinguistics is not an applied science, although its results are applied in practice—as, e.g., in the practice of teaching.

The emergence and development (against the background of the solution of production tasks) of applied human knowledge has been less studied in the methodology of science than the applied aspects of, say, natural science or mathematics. This applies to human knowledge both scientific and artistic.

Meantime, premises for the study of the fundamental and applied aspects of human knowledge have already evolved in the shape of a well-developed apparatus of the methodology of scientific knowledge and certain data accumulated in human knowledge. With reference to the human sciences, mention must be made first of all of such advanced scientific disciplines as linguistics and psychology.

It has been pointed out above that these sciences are *human* sciences because of the significant share in them of human knowledge proper that has not yet become separated out, and because of their relatedness to objects that are specifically human at the present stage in the development of culture. Considering the tendencies of development of these disciplines, especially the evolution of their applied aspects at a time of the scientific and technological revolution, these may be said to have remained human sciences for the present, although that fact is now questioned—for various reasons (33; 40). The knowledge of the human sciences proper in their applied aspects that have not yet become separate from the body of the base disciplines, is practically indistinguishable from the rest of knowledge in these disciplines. We therefore use the "classical" human sciences like linguistics and psychology to characterise the specificity of applied studies in human knowledge.

Applied linguistics as one of the two primary aspects of linguistics (along with fundamental studies) (185), uses linguistic knowledge for the solution of various practical tasks (32; 185). Consideration of the specificity of these tasks shows them to be production problems and questions: establishing mutual understanding in the man-machine system, speech control of production processes; automatic processing and classification of speech information, of technological documents, etc. At the same time linguistic knowledge can be applied, e.g., in historical science—for establishing the areas of settlement of different peoples and other problems where linguistics is not used in its applied aspects. In this and similar questions it would be more justified to speak of interdisciplinary rather than applied studies.

Applied aspects of psychology were in fact touched upon above, in the section on technical knowledge, where we discussed the methodological problems of ergonomics. The industrial character of the tasks being tackled in applied psychological studies becomes obvious in comparison with the applied aspects of other human sciences. The involvement of psychology in the solution of pressing production tasks in revealing the hidden potential for increased labour productivity, in designing and construction of technology, improving the systems of management of the national economy, etc., is so great that it led to the emergence of applied psychological disciplines—labour psychology, the psychology of management (organisational psychology), etc. Moreover, some applied psychological disciplines not only acquired the status of independent sciences but also developed into a basically new form of scientific knowledge. That applies, first of all, to human engineering or ergonomics as a non-classical scientific-technological discipline.

A considerable body of data has accumulated in the applied aspects of artistic cognition and art. The practice of decorative arts and industrial design will have to be interpreted in the light of epistemology. The controversy about pure art and art for art's sake in aesthetics and art studies in the broad sense has raged for quite a long time. The great creative potential of epistemology and methodology of science may and must be used to study the nature of applied art and applied artistic knowledge. In this way, the pure art situation can be elucidated, so to speak, by the rule of contraries. Summing up the practice of applied art and design in all its forms and directions, applied artistic knowledge (applied art) may be said to be the application of artistic knowledge for utilitarian purposes in everyday life and the industrial sphere.

Human-scientific knowledge. The problem of the human sciences is not solved once and for all by giving a list of disciplines or fragments of knowledge. Essentially it is the problem of adequately studying the sphere of the specifically human. The latter is highly mobile, which conditions the instability of the boundaries of the human sciences and human knowledge in general. Any science may be involved, through its fragment or method, in the study of this domain, and this is reason enough for describing it as a human science.

It does not follow, however, that any sort of arbitrariness

is permissible, still less necessary, in the methodological analysis of the human sciences. On the contrary, now more than ever, the human sciences are in need of clarity, rigorousness, and precision of the methodological instruments of their study.

In particular, the need has become apparent in the development of epistemological studies of knowledge in the human sciences for a final overcoming of interpretations of the human sciences which take shape outside the methodological analysis of scientific knowledge. We refer here to a widely current approach to the human sciences—the practice of "clarifying" the status of the human sciences in the curricula of the human knowledge departments of higher educational establishments or in various guidelines for the establishment of academic councils at colleges, institutes, universities, Academy institutes, etc. These decisions on the division of the human or any other disciplines are made necessary by various organisational elements in science, but far from determining or even replacing an epistemological or methodological analysis of scientific disciplines, they must rely on such an analysis.

An important element of epistemological explication of the human sciences is concretisation of the key concepts to which the specificity of knowledge in the human sciences is often reduced. For instance, there is the widely current view, going back at least to the neo-Kantians of the Baden school, that the human sciences are concerned, generally speaking, with values only. Naturally, a clear understanding of the specificity of the human sciences largely depends in this case on a clear interpretation of the concept of value. But, as O. Drobnitsky correctly noted, "the more universal, as an instrument of explication, the value concept becomes, the less we can explain the phenomenon of value itself. Intended to be the master key that opens all problems, it becomes a problem itself—a mysterious *x*. Whenever a philosopher despairs of finding an intelligent answer, he invokes the category of value. The problem is therefore not solved but merely labelled, 'These are values', as if that were enough to clear the issue. In the philosopher's mind, a value is a sort of limit to understanding" (29, 146-147).

It is generally recognised that in the 20th century the human sciences are moving from the empirical or descriptive level to the theoretical one. More specifically, the human disciplines may be said to be at the beginning of this transition.

As for the properly human sciences that separate from such disciplines as linguistics, psychology, culturology, the ensemble of the sciences of art (all those disciplines that may be called the "classical" human sciences), even the level of generally accepted views or conceptions has not been achieved here. This applies, first of all, to the psychology of personality, the psychology of emotions, text linguistics, linguostylistics, etc.

The inadequate theoretical level of the human sciences determines the ineffectiveness of their methodological studies, as manifested, among other things, in the absence of analyses of the theoretical schemata of the separate human sciences. In the natural and technical sciences, analysis of the formation of such schemata is carried out on the basis of a clear understanding of the organisation, methods and functions of this form of cognition. It is precisely at this stage that methodological reflection in the human sciences is at present. The epistemological study of knowledge in the human sciences is focused on problems of the theoretical and empirical level, the specificity of the methods and basic functions of the human sciences, criteria of scientificity, etc. Of course, reflecting on the methodological aspects of the concrete human disciplines (proceeding, explicitly, from methodological results), scientists attempt to attain the theoretical level. As a rule, however, numerous ideas and concepts of this sort find no general recognition. Evidence of this is found in the situation in individual psychology, psycholinguistics, linguostylistics, etc.

From the methodological point of view, the present-day situation in the human sciences is reminiscent of the situation in physics in the first quarter of the 20th century. Problems of physical knowledge of that time are generally known, well-studied and thoroughly described. We would like, however, to point out a certain aspect that is a recurrent feature in the development of scientific knowledge at the present stage; a discussion of this feature must not be seen as an attempt to reduce human knowledge to physico-mathematical natural science.

This aspect is well represented in Werner Heisenberg's book (142), in particular in the author's dialogues with Niels Bohr. The gist of Bohr's argument substantiating the new conception of atom structure, as Heisenberg remembers it, was as follows. In classical physics and science in general, to explain a new phenomenon meant to reduce it, through available concepts and instruments, to familiar phenomena and laws. This method and structure of thought were completely unacceptable in the description of the structure of the atom in nuclear physics, for one would have to resort to the concepts of classical physics, which are in this case inadequate and cannot cover all that occurs here.

To explain his thinking, Bohr resorted to an image that can symbolise the whole of non-classical science, creative thinking in general. In analysing the situation that emerged in the study of atom structure, Bohr said this to Heisenberg: "We are thus, to some extent, in the position of the sea-voyager who finds himself brought to a distant land where not only the living conditions are quite different from those of his native land but also the language of the people living there is quite strange to him. He has to reach an understanding, but he has no means of doing so. In a situation like this, a theory cannot 'explain' anything at all in the sense generally accepted in science. Here it is necessary to indicate certain connections and carefully move forward" (142, 62). Progress in science is always connected with breakthroughs into new spheres, and it is not enough to perceive the content of new ideas—as Heisenberg points out, "the structure of thinking must also change, if one wants to understand the new" (ibid., pp. 100-102).

The human sciences are at present in a situation like that—before a breakthrough into new territory. One of the first problems that arise in this connection is the problem of methods adequate to the solution of urgent theoretical and practical tasks. Method as an analogue of the fragment of reality toward which a concrete human science is directed and in the space of which it moves can develop into a theory thus taking the science to the theoretical level of cognition of its subject-matter.

The problems of the specificity of methods, which are now so urgent in the methodology of the human sciences, were already clearly realised a hundred years ago. Wilhelm Dilthey was one of the philosophers of the end of the 19th century who felt an acute need for a new methodology of the human sciences different from the natural-scientific methodology. In his methodological analysis of the human sciences ("sciences of the mind") Dilthey expressed the rational idea of "rebuilding the structure of thought" (Heisenberg) in the process of studying new objects different from the objects of natural science—individuality, personality, etc.

This idea was explicitly stated by Dilthey in his statements on descriptive psychology. In Niels Bohr's terms, Dilthey, having discovered certain real connections in the sphere of the psychical, began to move carefully toward definite knowledge—knowledge that could not be obtained by contemporary "explanatory psychology". That was why Dilthey, not unlike Bohr, chose as his bridgehead for a breakthrough into new territory, a critique of "explanatory psychology" or, more precisely, a critique of explanation in contemporary science.

In a book by John Stuart Mill published in Dilthey's time, we find the following definition of explanation oriented toward natural science: "Since explaining, in the scientific sense, means resolving an uniformity which is not a law of causation into the laws of causation from which it results,... if there do not exist any known laws which fulfil this requirement, we may feign or imagine some which would fulfil it; and this is making an hypothesis" (170, 322).

We need not evaluate here Dilthey's attitude to "explaining psychology" or his own theory of "understanding" psychology; we shall only point out that his rejection of mechanically transferring the procedure of "explaining"[16] from natural science into the sphere of psychology and the human sciences in general determined his search for new methods adequate to the subject-matter of human knowledge. In the process of that search, Dilthey formulated the following proposition: "The sciences of the mind must start from the most general concepts of the general theory of method and attain, through testing them against their particular objects, more definite procedures and principles within their domain, as the natural sciences have just done" (133, 143).

Gradually, Dilthey developed understanding as the principal method of the "sciences of the mind". Although the sphere of application of that method is practically unlimited, understanding in its highest forms is connected with cognition of the human individuality.

Dilthey failed to solve the key problems of the "sciences of the mind", yet in developing the method of understanding he plotted the "third path" toward the cognition of the human reality, one that was different both from natural-scientific and artistic knowledge. The content of Dilthey's method of understanding was the moulding of a structure of thinking different from natural-scientific thinking and aiming at universally valid scientific cognition of the human individuality and personality and of their manifestations in different spheres of cognition.

The familiar effectiveness of applying the procedures and methods of hermeneutics in the field of human knowledge and their extension of late to the natural sciences shows that concepts of understanding regarding it as a characteristic feature of cognition reflect a number of actual processes in the development of scientific cognition.

But can it be said that Dilthey, Betti or anyone else succeeded in working out or describing even the general features of a method of understanding? When speaking of understanding, we deliberately use the terms "characteristic" and "function" and avoid asserting that understanding constitutes a method. This is based on the conviction that understanding as a general scientific method or a method of human investigation has not yet taken shape or, at least, is not characterised in basic parameters.

Understanding as a complex issue of epistemology, linguistics, psychology, etc., is indissolubly linked with such fiercely discussed and little-studied questions as sense and dialogue. It will gain the status of a method only if the theory of understanding is developed in all its aspects. What we have at the present time are isolated interesting observations by M. Dummet on the link between understanding and sense, and by D. Follesdal on the hypothetical-deductive nature of the hermeneutic method.

In fundamental works and profound insights by Dilthey, Heidegger, Betti, Gadamer, Ricoeur and others, no attention was accorded to investigating the structure of sense, the comprehension of which constitutes understanding. Without an analysis of the component parts of sense, the elaboration of a theory, and thus the description of understanding as a method, are impossible; just as it is impossible to study the depths of matter using only, for example, the notion of the molecule.

Artistic knowledge. The second basic variety of the human knowledge is artistic cognition. The subject-matter of this type of cognition coincides with that of other human disciplines (individual psychology, the psychology of emotions, social psychology, etc.), but the method and instruments of coping with this subject-matter are different.

At the beginning of the present section the subject-matter of human knowledge was defined as man, the world of man, the human, etc. This view was described as the first step toward clarifying the subject-matter of human knowledge, including artistic knowledge. A more detailed study of the human element yields a more concrete definition of the subject-matter of artistic knowledge: it is defined as interpersonal individual relationships. Their cognition in art is mediated by the aesthetic ideal. The aesthetic ideal as the intersubjective subject-matter of art (38, 80) determines the most important aspects of artistic knowledge, such as the party spirit, and also the content and interconnection of elements in the artistic image as a system.

As a form of cognition, artistic knowledge differs from everyday consciousness (common sense) and scientific knowledge.

It is an "other-scientific form of knowledge". This expression (*inonauchnaya forma znaniya*) was aptly used by Sergei Averintsev to characterise symbology (1, 828); in using it, we have no intention of identifying artistic cognition with symbology but merely stress the element of *lack of opposition* between artistic and scientific knowledge, especially artistic knowledge and knowledge in the human sciences. Knowledge in art is "other-scientific" rather than "anti-scientific".

In the organisational principles, depth and modes of assimilation of the essence of reality, artistic cognition also differs from common sense. The principal structural unit of artistic knowledge is an artistic image. In the standard model of science, a scientific theory is a phenomenon that is of the same order as the artistic image. Studies in artistic cognition, especially those in aesthetics and literary criticism, often correlate the artistic image with concept in the sphere of scientific cognition as the principal "cell" of scientificity. This approach apparently does not take into account the level of development of the methodology of science at which the principal unit of scientific knowledge is a theory.

Although artistic knowledge is mostly oriented toward everyday consciousness, it represents an incomparably higher level of cognition with well-developed methods of idealisation, modelling, experiment, etc. On the analogy of the scientificity criteria, we can speak here of criteria of artistic cognition, which must not be confused with the criteria of artistic quality—a problem discussed in aesthetics, art criticism, etc. There are three such criteria, which run parallel with the scientific ones, and they are connected with the artistic image. The artistic image is a sort of theory in artistic knowledge. It is characterised by truth, systematicness, and intersubjectivity.

The specificity of these characteristics of the artistic image is such that the idea of artistic knowledge being scientific is instantly rejected. Proof of this is found in the truth of the artistic image, i.e., adequate reflection of reality in terms of an aesthetic ideal. The three above-mentioned criteria of scientificity specify the epistemological ideal of science, which in its turn serves as the standard of cognition in general. That is why art, being cognition (albeit artistic cognition), possesses these characteristics.

It is generally accepted in philosophical-epistemological studies that methodological reflection is a necessary condition of the development of science, of the generation of new ideas in it. This proposition is also valid for human knowledge—both in its scientific and artistic forms.

The purpose of methodological analysis in any sphere of knowledge is breaking out of the automatism of perception and solution of cognitive tasks. In the methodology of scientific cognition that is achieved through the realisation of the basic methods, procedures, goals, orientations of science, etc. In artistic cognition the instruments are different or even directly opposite, but the purpose is the same—the deepening of knowledge.

It is appropriate to mention in this connection the so-called effect or device of "estrangement" (*otstraneniye*) (25, 34-35). The methodological position of Victor Shklovsky and Bertolt Brecht demands that, in order to gain a deeper understanding of some phenomenon in the domain of artistic knowledge, that phenomenon must be seen as extraordinary or strange. In these two cases, the automatism of perception and solution of cognitive tasks and resultant deeper knowledge is attained in different ways. At the first stage of estrangement the object of cognition does not correspond to the customary, well-established norms and evaluations and is in this sense outside consciousness. However, the methodological instruments in the two cases differ precisely at the initial stages, for the ultimate goal of the estrangement effect is making the incomprehensible clear, introducing the subject-matter of artistic study into consciousness or, to be more precise, introducing it in the system of consciousness. The work of Bertolt Brecht and other outstanding artists provides evidence that artistic knowledge also shows a tendency toward transforming methodology into a mode of actual artistic practice.

Chapter 4

THE DIALECTICS OF DEVELOPING KNOWLEDGE

Non-Marxist philosophers are incapable of adequately reflecting the dialectics of the development of scientific knowledge. They cannot explain the phenomenon of science's steady progress toward objective knowledge increasingly conforming to reality and confirmed by socio-historical practice. This inability to provide acceptable explanations of the progress of science was one of the main causes of the crisis of Western "philosophy of science".

The present chapter will be devoted to an analysis of this complex and topical problematic.

4.1. THE CONCEPT AND PATHS OF SCIENTIFIC PROGRESS

The progress of science is its inherent movement from a less complete and precise truth to a more complete and precise one, a process that is conditioned and controlled by mankind's total socio-historical practice. The progress of science is the purposive, goal-directed movement of cognition toward more adequate, fundamental and universal conceptual forms all linked together by the unified advance from lack of knowledge to knowledge, by intellectual development.

The principles of progressive development of science can thus be represented by the pulsating knowledge model. The entire wealth of the dynamics of cognitive forms comprising science can be expressed by a modification of the two-plane structure whose favoured direction of movement and change is specified by the vectors of productiveness and criticism. The vector of productiveness or expansiveness stimulates expansion of the body of knowledge. The criticism vector stimulates careful analysis of accumulated knowledge, ousting out unsuccessful tests and unjustified elements, and leading to compression of knowledge. Under the influence of these vectors of research activity, knowledge is now expanded, now compressed. As a result, a "sediment" appears.

Let us use in this connection the concept of hard core—a suitable term for an adequate interpretation of this phenomenon.

Theoretical courses are regularly re-written after paradigmatic discoveries or scientific revolutions determining progress in the sphere of knowledge. That is quite understandable, as a theoretical course of some science embodies the most complete knowledge in the given area. This re-writing of courses permits a re-interpretation of the totality of accumulated knowledge from the standpoint of innovative knowledge, of the most informed front-line knowledge, so that certain corrections, clarifications, re-interpretations, etc., are introduced in the history of the concepts of a certain science.

Thus any science, in the intensive sense, can be actually identified with a currently accepted fundamental theoretical course. If we analyse the content of all theoretical courses in a certain science re-written throughout its history, we shall discover that some elements in these courses remain invariant. These elements—the immutable, unproblematisable content constituting the "ideal course of the science"—is represented by the concept of hard core, which acts as the epistemologically ultimate basis for evaluating the degree of progress in the science.

In this case, the development of a science represented, as it were, ideally, is realised as the process of critical absorption of the truth in the "ideal course", while the development of real science, actually represented as a kind of multilevel cloud shrouding the "ideal science" and non-uniform in terms of adequacy (problematic and hypothetical), is realised as a process of necessary and often difficult evolution toward the "ideal".

The most fundamental premise of the development of science is mankind's aggregate socio-historical practice. This will be quite clear if we take account of two circumstances. Firstly, practice is a stimulator of cognitive progress, which follows from the conditions of the social functioning of science as a purposively organised institution serving the needs of society and of reproduction on a large scale. Secondly, practice is an instrument of substantiation of science, which follows from the possibility of verification, confirmation or rejection (of distinguishing between truth and falsehood) of the products of theoretical activity which is not self-sufficient in the sense of being substantiated *ex principio interno*.

Taking into consideration the fundamental mediating function of practice in relation to the development of science, we must not at the same time see it as the only premise of that process. Being a relatively autonomous sphere of the social superstructure, science has an independent logic of development and is subject to dimensions of change inherent in science alone. From this standpoint, whatever factors or stimuli may make an impact on science, it will not be receptive to them as long as they are not problematised and conceptually assimilated.

Life poses problems. Solving or eliminating these problems is the task of science. The most that the extra-scientific sphere can do is ensuring the conditions of productive development of science, creating the basis for its inner progress. The latter is confirmed, in particular, by the existence of real and significant problems which present-day science cannot eliminate. Thus practice has long posed the need for solving the energy, oncological, ecological and other problems for which, however, there is no solution in science in view of the lack of necessary conditions in it.

Taking this into account, let us discuss the internal premises and mechanisms of the development of knowledge.

The cognitive premises proper which ensure the progress of knowledge are polymorphism, inverseness, and incompleteness. Let us characterise all three.

Polymorphism. The concept of polymorphism indicates that the semantic resources of any scientific concept incorporated in a definite system of knowledge are never fully exhausted, and that there is always a possibility for an expansion of the conceptual sphere, determined by the existence of reserves, of a "free path" of the concept in question, a possibility that engenders a new semantic potential and consequently leads to the assertion of new systems of knowledge. The growth of knowledge is always connected with and accompanied by the discovery of new horizons of the basic concepts, the revelation of their fresh semantic strata, the formation of unprecedented combinations, and the discovery of mutual transitions and connections which were only present in potential form at the initial stage or, at any rate, could not be brought to light.

It means that the properties of the scientific language itself permit overcoming the feature of thinking which can be tentatively called ossification, which is connected, in particular, with the rigid formal logical organisation of the conceptual apparatus. It would, of course, be naive to assume that scientific concepts can be unorganised from the formal logical point of view; they must be sufficiently clarified—which is, properly speaking, an elementary condition of correct demonstration, one that rules out trivial logical errors like *ignoratio elenchi*, quadrupling of terms, etc. And yet the property of polymorphism of scientific language must be seen as a most remote premise of the growth of knowledge, signifying as it does a natural possibility of overcoming the formal requirements of non-ambiguity (in the sense of ossification rather than clarification). The following explanation seems appropriate in this connection: the higher the formal organisation of the language of theory, the greater its non-ambiguity and the less its capability for expansion, for being non-trivially self-expanded and thus involved in a description of an additional sphere of phenomena. Contrariwise, the greater its expansiveness, the lower its level of formal organisation.

Thus the dialectical contradiction, inherent in science, between non-ambiguity and polymorphism is a premise of the growth of knowledge. Indeed, "fuzzy and vague words with their rough edges, lack of clarity in the boundary lines dividing concepts, their diversity and variety—all this creates the possibility for violating the strictly deductive forms of thinking" (65, 133). In other words, scientific thinking naturally "must be logical enough, i.e., it must be based on deductive logic", but to be heuristic, "it must be built in such a way as to permit violations in the strict logic of a system of postulates or in the rules of deduction" (ibid.)—violations that later permit unfolding non-demonstrative modes of drawing conclusions (forms of thinking). For greater clarity, let us stress that in any logically correct thinking the rules of deduction cannot be "violated", otherwise science would become a collection of paralogisms. Referring here to "violations" in "the strict logic of a system of postulates or in the rules of deduction", we have in mind, just as the authors of the quote, merely a premise for progress in knowledge consisting in the fundamental ability of scientific concepts to change their content, which permits overcoming the rigid logically non-ambiguous structure of available knowledge, determining progressive conceptual shifts in it.

Thus pointing to the capability of scientific concepts for changing their content, the property of polymorphism characterises the development of knowledge in terms of its conceptual mobility. Progressive functioning of knowledge, realised as continual change in its content, is only possible through polymorphism, which introduces additional information in the available conceptual fund.

Inverseness. The inverseness of scientific objects (ideas, principles, objects) is their polyfunctionality, mutual reversibility, substitutability; in other words, their ability to act as the "proper" entities of qualitatively different and even epistemologically incommensurable systems of knowledge (cognitive connexts). Incorporation of an inverse object in a definite system of knowledge is a premise of its inner progress. Inverseness is based on the reinterpretation in scientific cognition of the semantic and operational status of objects in the framework of a certain closed cognitive context, a reinterpretation that leads to the establishment of new associative links moving these objects into unfamiliar conditions, which determines the possibility of progressive shifts in the problematic and conceptual fields of knowledge.

Let us cite a concrete example from the history of science to explain our thought.

Aristotle formulated the teleological principle (which was later elaborated by the Peripatetics and Thomists) that, through the Providence of the Creator, everything in nature is realised with a minimal expenditure of action. In 1744, Pierre Maupertuis, in a memoir sent to the Academy of Paris, borrowed that principle from metaphysics to explain certain optical and mechanical phenomena. Some time later he generalised that principle to embrace all movement, invoking a certain general principle of "least action in nature" asserting the minimal quantity of action. From this Maupertuis derived the laws of the lever and the impact of resilient bodies.

Maupertuis applied this principle to finite changes of velocities. Later Leonhard Euler generalised it to include continuous movements; unlike Maupertuis, he also connected it with the law of living forces: the sum of all the living forces in a body is the least of all possible sums.

Later Joseph Lagrange extended Euler's principle to include the material point in an arbitrary system of points, and applied it to the dynamics of the system. In the end, Karl Jacobi called the principle of least action the mother of analytical mechanics.

What is the dialectics of change in knowledge illustrated by this case?

At the first stage, there exists an idea that has no direct scientific meaning. At the second stage this idea is implanted in concrete knowledge, acquiring real intrascientific content (explanation of optical and mechanical phenomena, derivation of the lever law, and the law of impact of resilient bodies, carried out by Maupertuis). That content is fairly poor, and it is greatly tinged by the original interpretation depending on the specificity of the donor field.

Thus Maupertuis substantiated the physical application of this principle by the fact that it left the world constantly needing the omnipotence of the Creator, and was a necessary consequence of the wisest application of that omnipotence. At the third stage, the empirical field of the action of the principle is changed (Euler's innovations), while its general theoretical interpretation is retained and remains fairly immutable (cf. Jacobi reproaching Euler, d'Alembert and other scientists for a metaphysical evaluation of that principle). At the fourth stage, the filling of this principle, initially alien to the given science, by its 'native' content is completed. We know that at the end of his treatise *De isoperimetricis* Euler showed that $m\ s\ vds$ (where m is mass; v, velocity; s, path; and s, integral) has a minimum; this last is regarded by Lagrange as the properly physical content of the principle of least action, which he regards as a simple general conclusion from the laws of mechanics rather than as a metaphysical principle. That is precisely the way in which the idea from a donor field is implanted in the body of the accepter knowledge.

Inasmuch as the nearest inverse objects are language units or terms, the development of science is based on continual incorporation of terms and concepts from contiguous language spheres. The latter is made possible by paronymy, or the use of words of everyday language (any non-proper language of a certain system of knowledge is an everyday language in relation to its own language) with strictly fixed meanings which become intrascientific. Characteristic in this respect is, for instance, the history of the appearance in physics of the term "quark".

More precisely, inverse concepts (ideas or objects) are introduced in the following manner.

First, this process may rely on and stem from contiguous systems of knowledge. Thus the concept of tension was introduced into construction theory by Louis Navier, who borrowed it from electricity; Georg Cantor introduced the physical concept of power in mathematical set theory, etc.

Second, the introduction of inverse concepts may be based on a highly allegoric basis, and its genesis may be "irrational". Consider the history of the introduction of the term "quark". The concept of potential originally did not have a concrete physical meaning either, and was seen as a "working hypothesis" brought to physics from everyday language. Only after the establishment of a clear-cut categorial grid combining work, energy, conservation laws, etc., did it receive a physical interpretation.

Third, inverse concepts may be introduced in the form of using "traditional" objects or concepts whose meaningful and functional status is reinterpreted. Atomism may be taken as our example here.

From the positions of inverseness, progress in science is thus modelled as a process of acceptance or borrowing of "working abstractions"; of clarification of their meanings and senses; of final elimination of the "metaphoric" basis and imparting positive and intrascientific content to abstractions. Being, like polymorphism, a premise of the progress of science, inverseness ensures the conditions for restructuring the conceptual fund of knowledge. Polymorphism determines the mobility of systems of knowledge, whereas inverseness concretises the mode of the actualisation of that property. Any element of knowledge—scientific object, concept, principle, etc.—is mobile, among other reasons, because it is a dialectically contradictory fusion of reality and possibility. Reality is actual representation and expression of the object in the existing systems of knowledge. Possibility is the set of potential notions and expressions connected with the future involvement of the object (owing to its inherent inverseness) in the systems of knowledge that are next in turn. From this position, the possibility of progress in science consists in the unlimitedness and diversity of the objects of science, in the absence of monopoly of their "preferential" use in the framework of some conceptual construction given beforehand.

Incompleteness. Incompleteness, or lack of completeness, is a polysemantic factor which, if we ignore the formal-logical usage, means the following—as far as problems of development of knowledge are concerned.

Meaningfully interpreted, incompleteness means the absence of exhaustive information about reality studied by some area of knowledge, which is an objective premise of the possibility of progress in science; if that were not so, science would have neither room nor direction for progress. To clarify this, let us state that objective incompleteness of available theories determines progress in science, stimulating the subjective desire for achieving their completeness, which is expressed in the tendency to attain an epistemologically desirable exhaustive description, i.e., to formulate more powerful theories providing more information about reality.[1] In a more formal interpretation, incompleteness signifies absence of absolute closedness, or the possibility to invoke secondary considerations or assumptions that are not included in the propositions of systems of knowledge in their substantiation. The ideal of maximally complete (closed) scientific theory is unattainable, which follows from general philosophical, methodological and logico-mathematical arguments. Formal incompleteness of theories is a premise of their progress, making it possible to add further assumptions or propositions other than the accepted principles, postulates, axioms, etc., and to obtain in this way new theories.

The concrete processes in the development of knowledge determined by incompleteness are universalisation, integration and unification.

Universalisation. In discussing the progress of mathematics, Jean Dieudonné stresses that true progress is mostly determined by a deeper understanding of the phenomena under study, which is usually due to their incorporation in a broader framework. Research experiences have shown this to be true not only of mathematics but also of science in general. It is therefore important to make these two points: progress in science consists in deepening understanding, and deeper understanding is achieved through incorporation of previous knowledge "in a broader framework".

First of all, what does "deeper understanding" mean? L. I. Mandelshtam speaks in this connection of two phases in understanding. At the first phase, "you have studied a certain problem and seem to know everything you may need, but you cannot yet answer independently a new question pertaining to the field in question. In the second phase, a general picture and a clear understanding of all the connections emerge" (56, 10). Progress in science should be connected precisely with this degree of understanding. The essential point here is that the second phase of understanding comes only with knowledge that is deeper and more adequate than knowledge at the first stage, which ultimately solves the questions arising in the given domain.

For example, progress in the development of classical mechanics linked with the ultraviolet catastrophe consisted in the deepening of the general picture of the nature of atomic phenomena, of physical reality in general. That deepening involved a revision of a whole series of classical and semi-classical propositions: the principle of continuity of physical action and energy (introduction of the quantum postulate); Bohr's atomic model and the concept of electron orbits (introduction of the stationary state concept); the dynamic variable concept (introduction of new mathematical objects satisfying the condition of non-commutative multiplication); dynamic causality (substantiation of the objectiveness of statistical laws); the idea of commutativeness of all the physical characteristics of an object (introduction of the principles of complementarity and indeterminacy); the concept of state specified by a point in the phase space of a system (representation of state by a vector in infinite-dimensional Hilbertian space). All this taken as a whole led to the emergence of quantum mechanics and concomitant deeper understanding of physical reality, of the nature of matter as such. This apparently determined not only the solution of the crisis in the theory of the absolutely black body but also stimulated general progress in physical science on the basis of a clear understanding of the essence of the object under study (the idea of quantification of physical action, the concept of discreteness of the process of interaction between radiation and matter, etc.).

Let us explain the proposition that deeper knowledge is achieved through incorporation of previous knowledge in a broader context.

The following law in the development of knowledge has long been established: that which appears as a more or less accidental fixation of an object at the empirical level in a previous theory (T_1), reappears as a necessarily derivable consequence at the theoretical level in a subsequent cognate theory (T_2). Thus Lorentz's ether at rest theory explained the negative result of the Michelson-Morley experiment on the basis of the hypothesis of the reduction of the linear extent of bodies in the direction of movement, the theory thus becoming an *a tergo* description of empirical facts. On the other hand, in the special theory of relativity the negative result of the Michelson-Morley experiment is, so to speak, planned or extrapolated at the theoretical level thanks to a new spatio-temporal picture. The same should be said of the red shift phenomenon, which is merely stated in the special theory of relativity but adequately explained in the general theory of relativity. Thus on a meaningful plain a "broader framework" signifies nothing other than a greater degree of development of T_2, its depth, which is a result of clarification of T_1, which permits theoretical prediction and derivation in T_2 of a fact that was empirically stated in T_1 but was not fully understood in it. In formal terms, a "broader framework" signifies the availability of a new theory from which the previous one is derived as a limiting case.

Integration. The tendency toward overcoming incompleteness is also manifested in integration, which, along with differentiation, is a powerful instrument of the development and formation of science. Sciences emerge as a result of their cognitive self-determination: a complex of problems (the subject-matter) emerges, through differentiation and integration, from a certain primitive knowledge of which a historically "limiting" expression is inarticulate archaic protoknowledge; this complex of problems requires special methodological study, and that gives rise to derivative science. Characteristic in this respect are the so-called synthetic sciences which emerge where there is a need for the description of a certain object in terms of many rather than one system of knowledge. Of this nature is, say, a scientifically substantiated description of the biosphere which cannot be presented in the language of any one science, for the essence of the biosphere is such that its scientific description is only attainable through simultaneous application of the concepts of many sciences. In this way integration processes bring about synthetic unification of previously unconnected branches or areas of science, which leads to a deepening of the conceptions of the nature of the phenomena under study. Thus Maxwell's electrodynamics, unifying a series of empirical laws accumulated in the autonomous and isolated theories of electricity, magnetism and optics, deepened and optimised physical knowledge. With the emergence of Maxwell's electrodynamics, electrostatic and electrodynamic phenomena, the phenomenon of electromagnetic induction, the laws of Coulomb, Ohm, and Ampère, many optical phenomena, etc., were given a profound theoretical interpretation from a unified position, which is a certain indication of scientific progress. Synthetic tendencies in science manifested themselves in the emergence of the special theory of relativity (a synthesis of mechanics and electrodynamics) and quantum mechanics (a synthesis of corpuscular and wave mechanics); a unified field theory is worked out through synthesis of the general theory of relativity and quantum mechanics.

Unification. One of the stimuli for the development of science is the tendency which is expressed, for instance, in Leibnitz's minimax requirement asserting the epistemological desirability of theoretical expression of a maximal number of essences through a minimal number of independent assumptions. The evolution of knowledge may be said to comply with this requirement. It is from this angle that one ought to consider the numerous restructurings and reorganisations of theories (including those based on axiomatisation and formalisation) indicative of the search for the optimal mode of their construction.

Inasmuch as one theory differs from another, as Einstein put it, mostly in the choice of its foundation stones, that is, the irreducible basic concepts out of which the structure is built, progress in science consists in choosing increasingly more fundamental stones capable of carrying increasing loads despite their diminishing numbers. Hence Planck's radical call for "unifying the particoloured variety of physical phenomena in a unitary system and even possibly in a single formula" (178, 6).

Thus incompleteness is a premise of progress in knowledge which characterises both its meaningful aspect, connected with the need for a transition from less fundamental (in terms of breadth and depth) knowledge to increasingly more fundamental knowledge, and its formal aspect connected with the need for the restructuring and reorganisation of available theories with a view to maximal realisation of the requirement of closedness.

The process of production of new knowledge is determined by three factors: non-logical forms of reasoning (productive capacity of the imagination, intuition) which act as the generative structures of knowledge; logical forms of reasoning embodied in symbolic articulations (categories, propositions, syllogisms, abstraction, theoretisation, etc.), which act as the ordering structures of knowledge; and the socio-historical process of assimilation, taken to mean the introduction of knowlledge in society's practice, i.e., its transformation into material and non-material culture.

One must not underestimate the significance of the analysis of the way in which our reasoning (which mostly occurs below the threshold of consciousness) is subsequently moulded in categorial structures and "conscious forms", and that which is produced by individual scientists becomes generally valid knowledge. This analysis is quite capable of discovering certain general rules of associative thinking (consider heuristics, for example) which cannot be definitively studied by the means of logic.

There are two paths of the development of knowledge—the evolutionary (extensive) and the revolutionary (intensive).

Evolutionary development, which does not assume any radical renovation of the theoretical fund of knowledge, consists in extending the sphere of application of available theories to cover new phenomena of reality. This is achieved by the following procedures: derivation of consequences (the "routine" activity of scientists in the logico-methodological disciplines); adaptation of a general theory to the solution of specialist tasks by adding corresponding assumptions (point mechanics vs continuous medium mechanics); merging of a mathematical formalism with a concrete theory, leading to the formation of new notions and systems of concepts (cf., e.g., the apparatus of mathematical statistics and probability theory—thermodynamics—statistical physics); development of theory through introducing new suggestions (improvements on heliocentrism by Kepler); a search for models and semantic (empirical) interpretation of theories (the work of Eugenio Beltrami, Klein, Poincaré and other scientists on the interpretation of non-Euclidian geometries).

In the most general but precise sense, the specific features of evolutionary development of scientific knowledge are described by the mechanism of homogeneous and heterogeneous reduction.[2] Homogeneous reduction, in which the theory subjected to reduction and the theory to which the former theory is reduced are both expressed in the same language, is a version of a fairly trivial expansion of the fund of knowledge. In this case, one of the theories (the expanded one) is the basis from which the cognate theory follows as a consequence. The general scheme of progress is in this case as follows: $T_1 H \rightarrow T_2$ (where T_1 is a basic theory and T_2, a derived one; H is a set of additional hypotheses expanding T_1). Examples here are classical mechanics and the more special classical theory of gravity derifed therefrom.

Heterogeneous reduction covers the relations between qualitatively different theories formulated in different languages. Inasmuch as the derived theory T_2 comprises in this case a set of conceptual units which are simply absent in the arsenal of T_1, the operation of trivial syntactic inclusion of T_2 in T_1 proves to be impossible. Consequently, the general scheme of progress in this case is as follows: $T_1 \cdot RS \rightarrow T_2$ (where R is a set of transitional propositions, and S, a set of semantic assumptions). This case can be illustrated by the synthesis of electrostatics and theoretical mechanics and the development of a generalised mathematical theory of potential. On the one hand, there existed a purely physical range of ideas and modes of studying electricity, as represented, say, by the work of Coulomb. On the other hand, there were purely mathematical analytical methods of theoretical mechanics developed by Euler, Lagrange, Laplace, and others. Poisson combined these two areas in handling the Coulomb law in mathematical terms and using it to solve "the mathematical task of the distribution of charge over different conductors and systems of conductors" (28, 296).

From the logical standpoint, a fundamental characteristic of evolutionary development is the existence of deductive links between a basic and a derived theories. The latter must be understood, in the first place, in the sense that constructing a derived theory (T_2) always has for a necessary condition the truth of the basic theory (T_1), for if the results of T_1 are true, the results of T_2 cannot be false.

Where the deductive links between T_1 and T_2 are violated, the development of science may be said to be revolutionary.

Revolutionary development of science, assuming a significant renovation and modification of its conceptual arsenal, consists in the deepening of previous notions of the essence of studied phenomena. A revolution in science is always a form of resolution of a contradiction between the heuristic potential of available knowledge and the factual material which must be interpreted in its framework. What are the conditions of revolutions in science?

Two causes may be pointed out as the most general ones. The first cause is the non-predicativeness of knowledge, which consists in any evolutionary development being accompanied ty the restructuring of logical, intrascientific and philosophical foundations of knowledge, which ultimately exhaust the immanent potential of self-development, causing the need for qualitative, revolutionary transformations. The second cause is the impossibility of infinite assimilation of new empirical facts by the available theory. Any theory assumes the assimilation of a definite body of empirical data. At a definite stage, however, saturation sets in, and from that point on a theory can no longer explain, describe, or predict empirical data. To recover the disappearing balance between theory and empirical data, additional hypotheses and assumptions are introduced in the theoretical apparatus which lead to agreement between theory and facts. However, assimilation of empirical counter-examples through *ad hoc* modifications in theory is not unlimited; it entails a loss of predictive potential by the theory in question, which becomes an *a tergo* description. Besides, the permanently increasing complexity of the apparatus of the theory leads to factual impossibility of its employment in current practical work, which demonstrates its inner imperfection. When this is manifested in an all-round reflection on available knowledge, the task is posed of its cardinal improvement, associated with the revolutionary mode of progress in science.

The premises of revolutions in science are (a) self-exhaustedness, the absence of the heuristic capacity of available knowledge to explain, describe and predict things; (b) increasingly artificial modifications of the apparatus of the theory, intended to adapt it to the solution of intratheoretical tasks; (c) contradictions, antinomies and other defects which discredit the traditional algorithms of the formulation and solution of problems.

These causes are not, however, sufficient for a revolution. A new idea must emerge, one that will indicate the direction of restructuring available knowledge. One must therefore be aware that a new theory, however rudimentary, must have evolved before we can discern antinomies in the facts contradicting an existing theory and thereby discrediting the tactics and strategy of research which it asserts.

A logical evaluation of revolutionary development of science should indicate a rupture in the deductive link between previous and subsequent knowledge. That means that a new theory cannot be derived as a logical consequence of an old one. The relations between a new theory and an old one are described in this case in terms of the correspondence principle rather than of the principle of deducibility.

The correspondence principle as a universal methodological principle solves the following problem. Any new theory (T_2), however non-traditional, emerges as a result of resolving contradictions in an old theory (T_1), and in view of this inevitably uses certain elements of its problem categorial fund. It follows that the distancing of T_2 and T_1 is not absolute, as there is always a possibility for their cognitive comparison. It is the principle of correspondence that realises the requirement of continuity between new and old theories. As formulated by I. V. Kuznetsov, it reads: "Theories whose correctness was experimentally established for a certain group of phenomena are not rejected with the emergence of new theories but retain their significance for the former domain as a limiting form and a special case of the new theories. The conclusions of new theories in the domain in which an old classical theory obtained, become the conclusions of classical theory" (49, 8).

Taking this into account, it is necessary to give a critical evaluation of the idea of incommensurability widely current in Western philosophy of science. The proposition that theories are basically incommensurable is an artificial one. On this approach, the situation is presented in such a way as if mutually replaceable theories were worked out somewhere outside science. If we should assert, for instance, that relativistic and classic mechanics are absolutely incommensurable, the special theory of relativity has no right at all to be regarded as a subsequent physical theory emerging in the course of resolution of the crisis enveloping classical physics. It is not clear, then, how and when it emerged and what causes brought it into being.

The insistence on the incommensurability of mutually interchangeable theories also appears to be grossly exaggerated. Thus there are obvious connections between the conceptual apparatuses of relativist and classical mechanics, despite their non-identity. For example, although Newton's and Einstein's masses have different semantic and operational meanings, the interpretation of mass as a measure of inertia holds for both cases, as Einstein's mass does not become an object absolutely different from Newton's mass, etc. The doctrine of incommensurability of theories, which actually rejects the interconnectedness of previous and subsequent stages in the development of knowledge, is opposed in dialectical materialism by the Marxist theory of development of science in which development is interpreted as coherent continuous movement from the old to the new, accompanied by a deepening of previously elaborated conceptions and their reproduction at a qualitatively new level.

4.2. THEORIES OF SCIENTIFIC PROGRESS

Science is not static—it is dynamic, processual, changeable. That is an empirical fact; it is so obvious that it cannot be problematised, it cannot even be included, in principle, in the sphere of the problematic. The whole question, however, is one of suggesting a conceptualisation or constructing a methodological or historiographic model that would be an adequate reflection, a theoretical interpretation and expression of that fact. This may be said to be a sphere of the highly problematic, in which considerable difficulties as well as objects of true problematisations emerge.

Difficulties and problematisations appear from the very beginning, when an attempt is made to designate or clearly formulate the problem. For example, it is a well-known fact that "not one trend in bourgeois philosophy either interprets or analyses scientific cognition as a process of development" (27, 255).

Indeed, the process of the development of science or knowledge is a permanent growth of its content potential—instrumental, categorial, factological—which reflects and expresses the orientation of science at a fundamental goal, that of adequate reflection of the essence of things, of assimilation of the truth.

On the other hand, in the traditional Western philosophy this problem has not been formulated even, in this sense, for the core of this philosophy is invariable rejection of the fact that progress in the cognitive sphere constitutes dynamic movement from the relative to the absolute, from the hypothetical and problematic to the unconditional and certain, and thus from the nearer essences to the essences of higher orders. For confirmation we can cite Popper's probabilism, Kuhn's conventionalism, Feyerabend's anarchism, Toulmin's ecologism, none of which bring out the objective basis of the direction of progress, and the causes of consistent purposive movement realised by science from a less complete and precise knowledge to a more complete and precise one.

A profoundly scientific understanding of the problems of developing knowledge is given in the framework of the dialectical materialist concept of scientific progress, which took shape as a critical overcoming of cumulativist and anticumulativist models of the dynamics of knowledge in the Western philosophy of science.

Cumulativism. In cumulativism, the development of science is perceived as a linear quantitative self-expansion of aggregate knowledge through monotonous addition of new truths. How do these new truths emerge? In answer to this question, let us point out that the instrument of implementation of cumulativism is continuism, postulation of uninterrupted continuous movement from past through present to future states of science, where the transition from one state to another is conditioned by a natural permanent addition to original knowledge of next-in-line elements of the same genetic and epistemic provenance. Solving the problem of the source of the new from these positions, cumulativists hold the view that "the new is the transformed old", discovering a certain historical precursor of any new scientific idea—which enables them to regard the development of science as a laminar process free from crises and cataclysms.

The precursor of cumulativism was the epistemological model of the growth of knowledge worked out in the classical philosophy of Modern Times, which presented scientific progress as extensive incorporation of new truths in existing knowledge. That model, which retained its positions in methodological consciousness up to the 20th century, was actively supported by Pierre Duhem, who used new historical materials to substantiate it, and who actually banished the notion of scientific progress from a number of methodological concepts.

At present, after the experiences of the scientific revolutions of the 20th century, cumulativism as a mental attitude has been done away with. No one takes a serious view of the cumulativist model of scientific progress, which presents the latter as a quantitative accumulation of absolutely true units of knowledge and sees its motive force as the orientation toward achieving an increasing level of systematisation and growing precision of measurement.

A critical evaluation of cumulativism as a methodological doctrine reveals the following shortcomings inherent it it.

(a) Cumulativists are incapable of introducing the relation of progressive change in the sense of a non-trivial conceptual shift in the present in relation to the past, or in the future in relation to the present.

(b) Contrary to the tenets of cumulativism, the realities of science indicate that the degree of influence of past knowledge is inversely proportional to the originality and creative abilities of scientists.

(c) Characteristic of cumulativism is an unacceptable rejection of the specificity of previous stages in the development (or history) of science compared to subsequent ones, which gives rise to an uncritical modernisation of the past and brings cumulativism, as a methodology, to discredited presentism and actualism.

Anticumulativism. Unlike cumulativists, anticumulativists believe the motive force of progress in science to be scientific revolutions—radical shifts and irreversible, leaplike transitions from certain phases in the history of science to other, qualitatively different, phases. Anticumulativism is realised in discontinuism, which rejects the universality of continuity and insists on the discontinuous "explosionlike" character of the development of knowledge. Anticumulativism describes fairly adequately such "epistemological leaps"—which are hard to explain from the positions of cumulativism—as the Copernican revolution, the transition from Peripatetic mechanics to classical mechanics, replacement of the phlogiston theory of combustion by the oxygen theory, etc. At the same time exaggeration of the role of scientific revolutions and underestimation of the importance of evolutionary stages leads anticumulativism up to blind alley of vulgar catastrophism, logically culminating in the factually disproved theory of permanent revolution.

All this is avoided in the dialectical materialist theory of scientific progress based on the principles of reflection, historicism, practical determinedness of cognition, the unity of evolutionary and revolutionary changes, continuity of development, and of the theory of the dialectics of relative and absolute truth and contradictory nature of the cognitive process.

In the most general outline, the dialectical materialist interpretation of the essence of progressive development of science is as follows.

In reality, no theory functions in science eternally; this follows from the relativity of cognition, and from the fact that the heuristic potential of any theory intended for the conceptualisation of a rigidly fixed domain cannot be extrapolated, on adequate grounds, to other domains requiring conceptualisation. That is why the deep (mediated) cause of the need for transformation of a theory is its objective non-universality and self-exhaustedness, manifested sooner or later; in other words, its incompatibility with the growing demands of science and material production. The nearer (or immediate) cause of the need for progressive changes in science is the discovery of logical and factual contradictions in theories; these changes, implemented both through evolution and through revolution, provide the modes of sublating these contradictions, conditioning the transition to better forms of the organisation of knowledge. The materialist dialectical evaluation of these forms is founded on the assumption that each of the historically realised forms of the organisation of knowledge "is a stage in ascending to a concrete reproduction of objective reality in theoretical concepts" (27, 259). The need for theory transformation is determined in this sense by transition to a more perfect and adequate (concrete) knowledge. The dialectical materialist theory of change in knowledge, which interprets its essence in terms of step-by-step approximation of reality through accumulation of certain content in the course of dialectical transition from old theories to new ones, has none of the defects of the theories of scientific progress worked out in the Western philosophy of science and appears to be the only adequate theory of the development of knowledge.

A more detailed interpretation of the progress of science will yield the following schema of step-by-step formation of new knowledge.

A new theory emerges as a result of elimination of contradictions in an old one through the development of non-traditional heuristics. How does it come into being? In answering this question, scientists usually appeal to intuition as the principal generator of the new. There are sufficient grounds for this position. Indeed, a scientist's activity is to a considerable extent programmed—by various instructions, imperatives, regulations, recommendations, schemata of model principles of analysis, research projects, heuristic instructions, norms, standards, etc.

The whole of this arsenal determines and specifies the generally valid, "normal", standard conceptual and methodological rhythm of scientific activity. Quite clearly, this activity produces, so to say, the pre-planned new. Nothing fundamentally new can be created and produced here. Hence the correctness of the appeal to the intuitive, subconscious, non-discursive, irrational strata as tools for the production of the fundamentally new.

The problem of the emergence of new knowledge has largely remained unexplicated, since researchers mostly confined themselves to references to intuition in their discussions of the source of new ideas in science, and the task of logical reconstruction of intuitive acts was not solved (probably because it was self-contradictory).

To make some headway in the solution of the problem, let us consider this question: What is intuition as a source of the new?

In answering this question, the following facts have to be taken into account.

Intuitive activity is always based on a shortage of information needed for discursive-logical processing of knowledge.

Being responsible for the generation of the new, the psychical mutagenesis determining intuitive acts is implemented through re-combining the traces of impressions received from without; it is not controlled "by a conscious effort of will; only the results ... of activity are submitted to the judgement of consciousness" (90, 26).

The intuition underlying a discovery is not an accidental, supernatural or non-intelligible mutation of thought. Firstly, superconsciousness itself "carries out the primary selection of emergent recombinations and submits to consciousness only those that are marked by a certain probability of their correspondence to actual reality" (ibid.). In other words, ideas with an intuitive genealogy are, strictly speaking, "crazy" but not "mad". Secondly, the activity of superconsciousness is from the outset "channelled by the quality of the dominant need and the volume of previously accumulated knowledge" (ibid., 28). It follows that the time must be ripe for the discovery, and new ideas must be in the air.

The general methodological ideas on the non-accidental nature of discoveries—owing to the programmed character of psychical mutagenesis, the existence of goal-directing dominants in the form of the need for eliminating "hot spots", determination of superconsciousness by the available fund of knowledge and practical experience—are concretised in various models. One of them interprets discovery as a porism, i.e., an unpredictable, unplanned result that is not a direct object of research and thus not expected by the researcher, a result obtained as an intermediate corollary in solving a scientific task. Of this nature is, e.g., the discovery of imaginary numbers which, as Felix Klein points out, recurred again and again in various computations regardless of and sometimes wholly against the will of the mathematician carrying out these calculations; only gradually, as their usefulness became more and more apparent, did imaginary numbers become more and more widespread (154, 61).

The idea of porism is a heuristic one. It sweeps aside the unacceptable view that a discovery is an illogical and irrational act, a consequence of discursively incomprehensible, inexplicable inspiration. Indeed, "the usual mode of reasoning is something like this: if a scientific discovery were a logical consequence of available knowledge, it would be predictable and thus could not be unexpected. But as a scientific discovery is unpredictable and unexpected, it is illogical" (23, 114). Porism rejects this type of argument; "unpredictability" and "unexpectedness" as a result of non-formalisability of creative activity can no longer "be arguments in favour of illogicality or irrationality"; "arguments against the 'logicality' of scientific discoveries also collapse" (ibid.).

Another model specifies the details of the mechanism of recombining the new out of the details of available knowledge, of the entire actual experiences. As the first step in the solution of the problem, two things must be opposed to each other: science and its immediate creator, the scientist. It must be borne in mind that the scientist does not always act in a way prescribed by science (by the scientific community). Inasmuch as scientific thinking is standardised and controlled by rigid model programmes, which substantially constrain the activity of the scientist, his deviation from these model programmes is, even on the purely formal side, a premise for the emergence of the new.[3] From the meaningful point of view, an explanation of the phenomenon of the emergence of the new comes with the realisation of the fact that in reality, the scientist works in several rather than one research programme (in view of his educational, professional, etc., polyfunctionality), whose intersections are the points of crystallisation of the new. Thus Louis Pasteur's refutation of the theory of self-generation, which made a noticeable impact on the progress of biological science, was only made possible by his training in physics, which helped him to sterilise more carefully the solutions he used, etc.

In terms of this model, the source of the new in science thus lies in the scientist using, in a certain research field, programmes and principles of analysis which have not yet asserted themselves in it as standard ones; more concretely, this source is believed to lie in the hybridisation of research programmes (86).

These models seem to throw some light on the dynamics of the formation of new knowledge from the very first stages in this process (though it would be premature to say that the problem has been fully solved); in itself, this inspires hope; at least, the problem is not relegated entirely to the field of psychology through bootless references to the intuitive and the subconscious.

After the formation of a new theory, which emerges as a result of the resolution of contradictions in an old one, the important and difficult process of its objectification begins. The epistemological content of this process lies in ascribing an ontological interpretation to the theory, its identification with a certain fragment of reality and correlation with the existing picture of the world. The process of the objectification of the theory can now take either of the following two routes.

A. The result of the objectification of a new theory corresponding to its interpretation (which, properly speaking, figures as the object of theoretical knowledge) is not incompatible with the existing interpretation of the phenomena of this class provided by the old theory. We observe this situation in the case of objectification of theories in the framework of the evolutionary path of the development of knowledge, e.g., as a result of proliferation or modification of dominant research programmes which do not result in a crisis of "normal" science.

B. The result of the objectification of a new theory proves to be incompatible with the traditional interpretation of events provided by the old theory. This situation is observed in cases of objectification of theories in the framework of the revolutionary path of development of knowledge, e.g., as a result of a radical rejection of dominant research programmes in connection with a crisis in "normal" science. The following situations may arise here.

(1) An attempt is made at combining, logically, the new and the old interpretation. It is realised as a tendency toward objectifying the content of a new theory in the framework of an old world picture. Tycho Brahe's astronomical system can be cited here as an illustration; clearly realising the intrascientific advantages of heliocentrism as compared to geocentrism, he nevertheless did not dare, for worldview considerations, to accept the former fully and finally, and revived the eclectic model of Heraclides Ponticus.

(2) An attempt is made to present the new interpretation as a temporary phenomenon on the whole unacceptable to science which must step down and give way to good old concepts, and the sooner the better. Of this nature is, e.g., the tendency to present quantum mechanics as an incomplete phenomenological theory requiring a reformulation in a dynamic determinist language.

(3) An attempt is made to show the inconsistency and untenability of a new interpretation with a view to excluding it from the domain of science. This can be illustrated by the well-known attempts to show the paradoxical nature of the conceptions about the discreteness of energy states of atomic objects from the positions of classical physics.

(4) The following case, probably the most interesting, is an inevitable culmination of the previous ones. After the final acceptance of the new and downfall of the old interpretation, an attempt is made at elucidating their mutual relations on the basis of the principle of correspondence. The ordering of existing interpretations on this principle, ultimately signifying (a) the establishment of the unity of knowledge about objects and (b) the deepening of theoretical notions about the phenomena of reality studied in knowledge, constitutes an objective basis for the qualitative progress of science expressed by the formula "toward what we want to know" and not, as Kuhn believes, "from what we know". The former, as it is easy to see, enables us to reject the destructive conviction that "sciences (theories) are simply different"— a view that denies the possibility of progressive changes in the sphere of knowledge, and to adopt the constructive conviction that "sciences (theories) can be more perfect", which permits an adequate basis for the theory of scientific progress.

The following can be presented as a general acceptable schema modelling the growth of a theory. The emergence of new empirical data, or the demand for semantic and logical optimisation of knowledge make it necessary to modify existing knowledge. From the genetic viewpoint, these modifications come into being either as porisms or as consequences of the realisation of the productive and creative intuitive potential of cognition on the basis of recombination of elements of available experiences (in particular, through hybridisation of accepted research programmes).

The innovations resulting from modifications of knowledge take shape and are organised as theories from which experimentally verifiable consequences are logically deduced; if they agree with experimentally verifiable state of affairs, theories are confirmed and introduced into practice. Further progress of theories is determined by the activity of their logical, operational. experimental, semantic, etc., perfection in accordance with the familiar criteria of internal perfection and external justification.

In the long run, as the theoretical and factual bases of knowledge coincide only partially, this natural cycle of science is reproduced again and again.

In relation to the basic unit of science, a scientific theory, which is not a direct and primitive coagulation of observations or empirical data, this schema can be presented in the following formula:

$$E - I - T_p - T_c - E^1 \quad (1)$$

where E is empirical provocative facts; I, intuition; T_p, theoretical principles; T_c, empirically verifiable consequences from the theory, E^1, the field of possible empirical confirmations of theory.

With due consideration for the expansion of the empirical basis of theory, the dynamics of new cycles is described, accordingly, by the formula

$$E_i - I - T_{pi} - T_{ci} - EE_i^l \quad (2)$$

which demonstrates the spiral-like development of science in accordance with its dialectical nature (110).

E in (1) and E_i in (2) differ from E^1 and E_i^1 respectively both intensionally and extensionally. E *and* E_i are anomalies from the positions of an old theory, which cannot assimilate them, and "provocative factors" of a new theory, which is as yet non-existent, whereas E^1 and E_i^1 are logically derived from the theory as programmed consequences. E^1 and E_i^1 include a class of predictable theories of potential verificators, and are therefore broader than E *and* E_i.

The safety margin of the development of a theory, its heuristic quality and non-triviality are not unilimited; as the theory's conceptual charge is exhausted (as indicated by the inevitable deformation of inner perfection and external justification), it gives way to its successor. This explains the rotation of the cycles of science.

The diachronic cross-section of the problem provides a formula of progress somewhat different from the above—the one obtained for the typological cross-section. Inasmuch as the conceptual (B_c) and empirical (B_e) bases are not in a one-to-one relation (which signifies the non-deducibility of B_c from B_e and non-reducibility of B_c to B_e), the genealogy of new knowledge can be presented in the following formula:

$$\bigvee \quad E - J - \begin{matrix} T_{p_1} \rightarrow T_{c_1} \\ T_{p_2} \rightarrow T_{c_2} \\ \vdots \quad\quad \vdots \\ T_{p_n} \rightarrow T_{c_n} \end{matrix} - E^1, \quad \bigvee$$

i.e., (E — I) entails a set of theoretical pictures ($\Sigma\ TP$)—alternative research programmes, theoretisations and descriptions. Interestingly, ($\Sigma\ TP$) does not entail a set of empirical pictures ($\Sigma\ EP$). This last assumption would correspond to the methodological doctrine of pantheorism, which exaggerates the fact of theoretical saturation of E and B_e. In actual fact the autonomous ingredients of ($\Sigma\ TP$) differ conceptually, being obtained through stratification and proliferation of B_c and having a unitary B_e fixed and presented in terms of a standard (for a local operation) operational basis B_o (measuring and computing devices, various apparatuses).

B_o and B_e are relatively stable and independent from B_c and can thus be instrumental in the testing and critical substantiation of the latter; this results in discarding elements of B_c yielding $\Sigma\ TP$ and in asserting some preferred variant which becomes generally accepted. This last point must be stressed for two reasons. First, we would like to show the untenability of both the "critically rationalist" and "historical" trends in post-positivism: the former sees B_e only as a mechanical tool for selecting irrationally generated elements of B_c, while the latter interprets B_e merely as an appendage of B_c devoid of any independent cognitive role. Second, our intention is to assert the dialectical materialist view that B_c and B_e, despite their independence from each other are not structures of the same order; B_e is more fundamental both in terms of genealogy of B_c, which may be considerable and not easily specifiable but necessarily experientally-empirical, and in terms of substantiation of B_c, which is ultimately always experimental.

CONCLUSION

Various aspects of the phenomenon of science have been studied in the works of Soviet and foreign Marxist philosophers. However, the need has arisen for constructing a general dialectical materialist theory of science covering the problems of its origin, functioning and development.

The present work attempts a consideration of certain general questions of the nature of scientific knowledge, of its historically and actually given forms, kinds and types. The starting point of this attempt, undertaken from the positions of the dialectical materialist methodology, is the assertion of the unity of scientific knowledge.

Despite all its diversity, science is unified by certain universal requirements which it satisfies. These include such normative regulators as the cause-and-effect typologisation and conceptualisation of phenomena, conformance to formal-logical canons of reasoning, experimental adaptedness of knowledge, theoretico-methodological monism, reproducibility of the results of cognition, their rational substantiation, etc.

These requirements form the concept of the world of science as a stable paradigm which embraces and permeates science as a whole and constitutes it as a special type of spiritual production.

None of this cancels, of course, the polymorphous quality of science—its diversity in terms of history, subject-matter, methods, etc. This circumstance determines the need for diversifying the normative regulators and prescriptions imposed on concrete types of knowledge, and the construction of specialised pictures of scientificity.

The present work is an essay in the construction of such general and special pictures.

NOTES

To Chapter 1

1 In the more compact symbolic notation, all this can be conveyed in the following formulas: if in the hard core of science we have

$$P\ (T,h/b) \to 1,$$

then in front-line science

$$S\ (T,h/b) \to -\log P\ (T,h/b) \text{ and}$$
$$C\ (T,h/b) \to 1 - P\ (T,h/b),$$

where P is probability; T, theory; h, hypothesis; S and C, indices of the degree of risk and novelty respectively.

2 See Chapter 3.

3 We have discussed this in considering the typology of phenomenalist-essentialist types of knowledge.

To Chapter 2

1 Cf. Aristotle's view that science emerged where men had leisure.

2 These could be "primary concretions, or corpuscles" (Boyle), "atoms" (Gassendi), etc.

3 Which is tantamount to the view of physical action as a function of material bodies, not of space itself.

4 The reference here is to a characteristic feature of classical scientists and not to the concrete principles of conservation, which continue to play a great role in non-classical science as well.

To Chapter 3

1 The same idea was expressed many years later by David Hilbert in connection with his critique of the Euclidian method of constructing geometry. "Despite the high pedagogical and heuristic value of the genetic method," he wrote, "the axiomatic method is preferable in the final presentation and definitive logical substantiation of our knowledge" (164, 242).

2 In our view, a sixth labour function should be added to these, namely the aesthetic one, which is different from the others and exists from the very inception of human labour. It consists, generally speaking, in lending aesthetic qualities to a product of labour, and is at present just as important as the other five.

3 The hypothesis, expressed in the literature, that *Homo habilis* was a dead-end branch (159) does not cancel the human nature and quality of *Homo habilis*. If this hypothesis is confirmed, the human characteristics of *Homo habilis* will be preserved in history in the same way as all the tribes and peoples that have disappeared for good remain in history as members of the human race.

4 Of the numerous works on the subject contributions by the following authors may be mentioned in this connection: J.R. Dixon (134), J.C. Jones (150), E.V. Krick (157), D. Meister (167), I.C. Wilson and

M.E. Wilson (194), W.T. Singleton (184), P.H. Hill (145), A. Hall (141), as well as articles in the collections (96-101).

5 It is appropriate and necessary to give explanations in connection with the given quotation from C. Mitcham and his and R. Mackey's remarks on page 7 of the introduction to (172). By technical knowledge (science) we mean knowledge about technics in the above-mentioned sense. The meanings and correlations of the terms "technology" and "technological knowledge" are different, although sometimes they intersect and even coincide with the meanings of the terms "technics" and "technical knowledge". This usage (and explanation) of the said concepts attests to the necessity of working out of one and the same conceptual apparatus and vocabulary in the epistemological studies of the knowledge embodied in the terms "technics" and "technology".

6 One tends to agree with Ladislav Tondl that "it is not quite clear what is to be understood by 'application' " (188, 12).

7 Other periodisations of technological knowledge (as, e.g., in 106; 124) are also possible. Some interesting ideas on the nature and development of technological knowledge were expressed by Koyré (18, 129-136).

8 This schema of the inception of technical sciences does not claim to be an exhaustive description of all the aspects of the formation of technical sciences or of the entire mechanism of their emergence. As we see it, it states one essential aspect of the emergence of the technological sciences. Other schemata of this process are possible and necessary, especially for the non-classical scientific-technological disciplines (22, 142).

9 This excerpt from the book by Kuznetsov—one of the best philosophical-biographical studies in Einstein's life and work—illustrates the traditional, psychological philosophico-aesthetic approach to the relationship between natural-scientific and human thinking. Human knowledge and natural science are considered in terms of creativity; motives of creativity, aesthetic and ethical evaluations of the results of creative work, etc., are studied, but the conceptual levels, the basic structures and methods of these levels in the two forms of cognition are not compared. Moreover, in Kuznetsov's book, interaction between human and natural-scientific knowledge excludes the borrowing of positions (see p. 601) and, apparently, of the conceptual structures of these forms of cognition in general.

10 This broad version of the concept of science can be regarded as a concretisation of the general model of scientificity which was suggested in Rozhansky's book (84, 5-11) and was used in the previous chapter in the study of the origin of science in general—its separation from the pre-scientific, mythological, etc., forms of spiritual production.

11 Realising the fact that the semantics of the term "schematisation" is highly laden in the history of philosophy and psychology, we nevertheless retain the term, meaning by it solely the creation of abstract objects.

12 Even the term "quadrivium" goes back to Boethius. It should be stressed that the arts of the quadrivium were interpreted in a way that appears unusual to the modern mind. Thus music and astronomy were included among the mathematical sciences, whereas geometry was more like geography in the modern sense of the term.

13 Out of considerations of style, we shall not use the category of "I" in concretising the "specifically human", as "I" forms the "intimate nucleus" and regulatory principle of personality (42, 31).

14 In psychology, just as in other human disciplines, there is no unitary theory of personality. For other interpretations of the structure of personality see (34; 73; 76; 77).

[15] For details of "I" and its structure see (43; 77).

[16] Major Soviet scientists have also pointed out the need for changing the principles of explanation in the solution of the fundamental philosophical-humanist problem—that of correlation of brain and consciousness (30, 82)

To Chapter 4

[1] Completeness is the ability of a theory to solve problems arising within the boundaries of its applicability.

[2] Aware of the drawbacks of these terms, we nevertheless use them in view of the research and lexical traditions in science. Later, as we shall see from the context, the term "reduction" will be used as a proper name rather than as a synonym of "reducing".

[3] The problem of a rational boundary between well-founded and unfounded deviation from the accepted research standards is never simple, as the functioning of science shows; it appears that it will never be given a "final" explication.

BIBLIOGRAPHY*

1. S.S. Averintsev, "Symbol", *A Short Literary Encyclopedia*, Moscow, 1972, Vol. 6
2. N.S. Avtonomova, *Philosophical Problems of Structural Analysis in the Human Sciences*, Moscow, 1977
3. I.S. Alekseyev, "Science", *A Philosophical Encyclopedic Dictionary*, Moscow, 1983
4. A.V. Akhutin, *A History of the Principles of Physical Experiment*, Moscow, 1976
5. L.B. Bazhenov, *The Structure and Functions of Natural Scientific Theory*, Moscow, 1978
6. L.M. Batkin, *Italian Humanists: The Style of Life and the Style of Thinking*, Moscow, 1978
7. M.M. Bakhtin, *The Aesthetics of Verbal Art*, Moscow, 1979
8. P. Bicilli, *Elements of Mediaeval Culture*, Odessa, 1919
9. I.I. Blekhman, A.D. Myshkis, A.T. Panovko, *Applied Mathematics*, Moscow, 1976
10. A.N. Bogolyubov, "Mathematics and the Technical Sciences", *Voprosy filosofii*, 1980, No. 2
11. A.N. Bogolyubov, "Fundamental and Applied Sciences (On the Genesis and Development of Applied Sciences)", *Historico-Mathematical Studies*, No. 20, Moscow, 1975
12. N.N. Bogolyubov, "The Significance of Fundamental Studies in Nuclear Physics", *Priroda*, 1979, No. 7
13. R.A. Budagov, *The History of Words in the History of Society*, Moscow, 1971
14. I.N. Burova, *The Paradoxes of Set Theory and Dialectics*, Moscow, 1976
15. L.E. Ventskovsky, *Philosophical Problems of the Development of Science*, Moscow, 1982
16. V.I. Vernadsky, B.L. Lichkov, *Correspondence Between V.I. Vernadsky and B.L. Lichkov*, Moscow, 1979
17. V.A. Vinokurov and K. A. Zuyev, "The Denumerable and the Nondenumerable in Computing Mathematics", *Voprosy filosofii*, 1982, No. 5
18. *In Search of a Theory of the Development of Science*, Moscow, 1982
19. P.P. Gaudenko, "Hermeneutics and the Crisis of the Bourgeois Cultural-Historical Tradition", *Voprosy literatury*, 1977, No. 5
20. *Epistemology in the System of the Philosophical Worldview*, Moscow, 1983

* References 1-115 are given according to the Russian alphabet.—*Tr.*

21. I.N. Golenishchev-Kutuzov, *The Mediaeval Latin Literature of Italy*, Moscow, 1972
22. V.G. Gorokhov, *A Methodological Analysis of Systems Engineering*, Moscow, 1982
23. B.S. Gryaznov, *Logic. Rationality. Creativity*, Moscow, 1982
24. L.D. Gudkov, *The Specificity of Human Knowledge (A Critique of the Views of Max Weber)*, Moscow, 1978
25. A.V. Gulyga, *Art in the Age of Science*, Moscow, 1978
26. *Dialectics in the Sciences of Nature and Man. The Unity and Diversity of the World, the Differentiation and Integration of Scientific Knowledge*, Moscow, 1983
27. *The Dialectics of Scientific Cognition*, Moscow, 1978
28. Ya. G. Dorfman, *The World History of Physics*, Moscow, 1974, Vol. 1
29. O.G. Drobnitsky, *The World of Objects Come to Life. The Problem of Value and Marxist Philosophy*, Moscow, 1967
30. N.P. Dubinin, *What Is Man*, Moscow, 1983
31. I.P. Yegorov, *On Mathematical Structures*, Moscow, 1976
32. V.A. Zvegintsev, *Theoretical and Applied Linguistics*, Moscow, 1968
33. V.A. Zvegintsev, *Language and Linguistic Theory*, Moscow, 1973
34. B.V. Zeigarnik, *The Theory of Personality in Psychology Abroad*, Moscow, 1982
35. B.I. Ivanov, V.V. Cheshev, *The Emergence and Development of the Technical Sciences*, Leningrad, 1977
36. *Ideals and Norms of Scientific Research*, Minsk, 1981
37. A. Yu . Ishlinsky, "The Interconnections between the Fundamental and Applied Sciences and Technology", *Philosophical Foundations of the Natural Sciences*, Moscow, 1976
38. A.T. Kalinkin, *Aesthetic Ideal, Art, Cognition*, Moscow, 1983
39. P.L. Kapitsa, *Experiment. Theory. Practice*, Moscow, 1981
40. B.M. Kedrov, "On the Methodological Problems of Psychology", *Psikhologicheskiy zhurnal*, 1982, Vol. 3, Nos. 5, 6
41. F. Kh. Kessidi, *From Myth to Logos*, Moscow, 1972
42. I.S. Kon, "The Problem of the Human 'I' in Psychology and Literature", *The Scientific and Technological Revolution and the Development of Artistic Creativity*, Leningrad, 1980
43. I.S. Kon, "The Category of 'I' in Psychology", *Psikhologicheskiy zhurnal*, 1981, Vol. 2, No. 3
44. I.Ya. Konfederatov, *James Watt, Inventor of the Steam Engine*, Moscow, 1969
45. A.M. Korshunov, *Reflection. Activity. Cognition*, Moscow, 1979
46. L.V. Krushinsky, *The Biological Basis of Intellectual Activity*, Moscow, 1977
47. B.G. Kuznetsov, *Einstein*, Moscow, 1979
48. I.V. Kuznetsov, *Selected Works on the Methodology of Physics*, Moscow, 1975
49. I.V. Kuznetsov, *The Principle of Correspondence in Modern Physics and Its Philosophical Significance*, Moscow, 1948
50. V.I. Kuzmin, "Different Trends in the Systems Approach and Their Epistemological Foundations", *Voprosy filosofii*, 1983, No. 3
51. V.I. Lenin, *Collected Works*, Progress Publishers, Moscow
52. A.F. Losev, *The History of Aesthetics in Antiquity. Aristotle and the Late Classical Epoch*, Moscow, 1975
53. A. R. Luria, "Psychology as a Historical Science", *History and Psychology*, Moscow, 1971
54. S. Ya. Lurye, *Archimedes*, Moscow-Leningrad, 1945

55. A.S. Maidanov, *The Process of Scientific Creativity*, Moscow, 1983
56. L.I. Mandelshtam, *The Complete Works*, Moscow, 1948-1950, Vols. 1-4
57. A.P. Mandryka, *The Evolution of Mechanics in Its Interconnections with Technics*, Leningrad, 1972.
58. A.P. Mandryka, *The Interconnections between Mechanics and Technics*, Leningrad, 1975
59a. F. Engels, *Dialectics of Nature*, Progress Publishers, Moscow, 1972
59b. K. Marx, F. Engels, *Collected Works*, Progress Publishers, Moscow,
59c. F. Engels, *Anti-Dühring*, Progress Publishers, Moscow, 1975
59d. K. Marx, *Capital*, Progress Publishers, Moscow, 1974
60. K. Marx, *A Contribution to the Critique of Political Economy*, Progress Publishers, Moscow, 1977
61. Yu.S. Meleshchenko, *Technics and the Laws of Its Development*, Leningrad, 1970
62. *Methodological Problems of Theoretical Natural Science*, Kiev, 1978
63. *Methodological Problems of Interaction Between Social, Natural and Technological Sciences*, Moscow, 1981
64. *Worldview and Methodological Problems of Scientific Cognition*, Moscow, 1983
65. V.V. Nalimov, Z.M. Mulchenko, "On the Question of Logico-Linguistic Analysis of Science", *Mathematisation of Scientific Knowledge*, Moscow, 1972
66. *Science and Culture*, Moscow, 1984
67. *African Paleolithic*, Moscow, 1977
68. K.K. Platonov, *The System of Psychology and the Theory of Reflection*, Moscow, 1982
69. M.V. Popovich, "The Verification of the Truth of a Theory", *The Logic of Scientific Research*, Moscow, 1965
70. Ye.V. Popov, *The Epistemological Essence of Technological Creativity*, Voronezh, 1977
71. B.F. Porshnev, *On the Beginnings of Human History*, Moscow, 1974
72. *Problems in the Explanation and Understanding of Scientific Cognition*, Moscow, 1982
73. *Problems in the Psychology of Personality*, Moscow, 1982
74. *Psychological Mechanisms of Goal-formation*, Moscow, 1977
75. *Psychological Studies in Intellectual Activity*, Moscow, 1979
76. *The Psychology of Individual Differences. Texts*, Moscow, 1982
77. *Personality Psychology. Texts*, Moscow, 1982
78. *The Psychology of Technological Creativity*, Moscow, 1973
79. B.S. Pyatnitsyn, V.S. Porus, "The Dialectical Aspects of Interconnections between Value and the Growth of Scientific Knowledge", *Voprosy filosofii*, 1979, No. 3
80. V.L. Rabinovich, *Alchemy as a Phenomenon of Mediaeval Culture*, Moscow, 1979
81. A.I. Rakitov, *Historical Cognition*, Moscow, 1982
82. I.I. Rezvitsky, *Personality. Individuality. Society*, Moscow, 1984
83. I.D. Rozhansky, *Anaxagoras*, Moscow, 1972
84. I. D. Rozhansky, *The Science of Antiquity*, Moscow, 1980
85. B.A. Rozenfeld, "Relativity Theory and Geometry", *Einstein and the Development of Physico-Mathematical Thought*, Moscow, 1962
86. M.A. Rozov, "The Paths of Scientific Discoveries", *Voprosy filosofii*, 1981, No. 8
87. V.M. Rusalov, *The Biological Foundations of Individual-Technological Differences*, Moscow, 1979
88. V.N. Sagatovsky, "The Principle of the Concreteness of Truth in the

System of Subject-Object Relations", *Filosofskiye nauki*, 1982, No. 5
89. S.A. Semyonov, "An Outline of the Development of the Material Culture and Economy of the Paleolithic", *The Sources of Humanity*, Moscow, 1964
90. P.V. Simonov, "The Unconscious in the Psyche: The Subconscious and the Superconscious", *Priroda*, 1983, No. 3
91. *Systems Studies: Methodological Problems. Yearbook 1983*, Moscow, 1983
92. O.M. Sichivitsa, *Complex Forms of the Integration of Science*, Moscow, 1983
93. V.S. Stepin, *The Formation of Scientific Theory*, Minsk, 1976
94. *Technics in Its Historical Development*, Vol. 1, Moscow, 1979
95. O.K. Tikhomirov, *The Psychology of Thinking*, Moscow, 1984
96. *VNIITE Papers. The Engineering Aesthetics Series.* The All-Union Research Institute for Engineering Aesthetics, No. 39, "The Function of a Thing as an Object of Study in Design", Moscow, 1982
97. *VNIITE Papers. The Engineering Aesthetics Series.* The All-Union Research Institute of Engineering Aesthetics, No. 36, "Problems in the Formation of Design Programmes", Moscow, 1982
98. *VNIITE Papers. The Engineering Aesthetics Series.* The All-Union Research Institute for Engineering Aesthetics. No. 23, "Construction, Function, the Artistic Image in Design", Moscow, 1980
99. *VNIITE Papers. The Engineering Aesthetics Series.* The All-Union Research Institute of Engineering Aesthetics, No. 22, "Theoretical and Methodological Problems of Artistic Designing of Complex Objects", Moscow, 1979
100. *VNIITE Papers. The Ergonomics Series.* The All-Union Research Institute of Engineering Aesthetics, No. 17, "Problems of Methodology in Ergonomics", Moscow, 1979
101. *VNIITE Papers. The Ergonomics Series.* The All-Union Research Institute of Engineering Aesthetics, No. 20, "Problems in the Methodology of Ergonomic Studies", Moscow, 1981
102. L.I. Uvarova, *Science as a Productive Force of Society*, Moscow, 1982
103. M.I. Uryson, "Darwin, Engels, and Some Problems of Anthropogenesis", *Sovetskaya etnografiya*, 1978, No. 3
104. M.I. Uryson, "The Sources of the Humankind in the Light of the Latest Data", *Voprosy istorii*, 1976, No. 1
105. K.E. Fabri, *Instrumental Actions of Animals*, Moscow, 1980
106. V.M. Figurovskaya, *Technological Knowledge. The Specific Features of Its Emergence and Functioning*, Novosibirsk, 1979
107. *Philosophical Questions of Technical Knowledge*, Moscow, 1984
108. *Philosophical Problems of Modern Chemistry*, Moscow, 1971
109. S.O. Khan-Magomedov, "Some Topical Problems in Design Theory", *Tekhnicheskaya estetika*, 1983, No. 2
110. V.S. Chernyak, "An Essay in Structural Analysis of the History of Science", *Voprosy filosofii*, 1983, No. 2
111. V.V. Cheshev, *Technical Knowledge as an Object of Methodological Analysis*, Tomsk, 1981
112. E.M. Chudinov, *The Nature of Scientific Truth*, Moscow, 1977
113. V. S. Shvyryov, *The Theoretical and the Empirical in Scientific Cognition*, Moscow, 1978
114. G.I. Shemenev, *Philosophy and the Technical Sciences*, Moscow, 1979
115. M.L. Shubas, *Engineering Thinking and the Scientific and Technological Progress*, Vilnius, 1982

116. R.L. Ackoff, *The Art of Problem Solving*. Accompanied by Ackoff's tables, New York, A. Wiley-Interscience Publication, John Wiley and Sons, 1978
117. *Aristotle's Metaphysics*. Bloomington-London, Indiana University Press, 1966
118. G. Boole, *Collected Logical Works*, Vol. II, Chicago and London, The Open Court Publishing Company, 1940
119. M. Born, *Experiment and Theory in Physics. Address Given to the Durham Philosophical Society and the Pure Science Society, King's College, at Newcastle-upon-Tyne on 21 May 1943*
120. M. Born. "Continuity, Determinism, and Reality", *Det Kongelige Danske Videnskabernes Selskab. Matematisk-fŷsiske Meddelelser*, bd. 30, No. 2. Copenhagen, 1955
121. N.Bourbaki, *Eléments d'histoire des mathématiques*, Paris, Hermann, 1960
122. G. Bugliarello and D.B. Doner, *The History and Philosophy of Technology*, Urbana, Chicago-London, University of Illinois Press, 1979
123. M. Bunge, "Technology as Applied Science", *Contributions to a Philosophy of Technology*, Dordrecht-Boston, D. Reidel Publishing Company, 1974
124. S.R. Carpenter, "Modes of Knowing and Technological Action", *Philosophy Today*, Vol. 18, Summer 1974, No. 2
125. V.G. Childe, "Documents in the Prehistory of Science (II)", *Cahiers d'Histoire Mondiale*, 1954, Vol. II, No. 1
126. V.G. Childe, *Society and Knowledge*, New York, Harper and Brothers Company, 1956
127. A. Church, "Mathematics and Logic", *Logic, Methodology and Philosophy of Science*, Stanford, Stanford University Press, 1962
128. A. Church, *Introduction to Mathematical Logic*, Vol. 1, Princeton University Press, 1956
129. A. Clairaut, *Théorie de la figure de la terre*, Paris, A. Hermann et Fils, 1909
130. J.D. Clark, *The Prehistory of Africa*, London, Thames and Hudson, 1970
131. A.C. Crombie, *Robert Grosseteste and the Origins of Experimental Science*, 1100-1700, Oxford, Clarendon Press, 1953
132. R. Descartes, *Discours de la méthode*, Paris, Ernest Flammarion, o.d.
133. W. Dilthey, *Gesammelte Schriften*, Bd. V, Leipzig Verlag von B.G. Teubner, 1924
134. J.R. Dixon, *Design Engineering: Inventiveness, Analysis and Decision Making*, New York, McGraw-Hill Book Company, 1966
135. P. Duhem, *La Theorie Physique. Son Objet—Sa Structure*, Paris, 1914
136. A. Einstein, "On the Method of Theoretical Physics", *Ideas and Opinions*, London, Alvin Redman Limited, 1956
137. A. Einstein, "Physics and Reality", *Ideas and Opinions*, London, Alvin Redman Limited, 1956
138. R. Feynman, *The Character of Physical Law*, London, Cox and Wyman Ltd., 1965
139. A. Fraenkel, Y. Bar-Hillel, *Foundations of Set Theory*, Amsterdam, North-Holland Publishing Company, 1958
140. A. Gellius, *The Attic Nights*, Vol. II, London, William Heinemann, MCMXXVII
141. A. Hall, *A Methodology for Systems Engineering*, Princeton, D. van Nostrand Company, 1965
142. W. Heisenberg, *Der Teil und das Ganze. Gespräche im Umkreis der Atomphysik*, München, R. Piper and Company Verlag, 1969
143. C.G. Hempel, *Aspects of Scientific Explanation*, New York-London, 1965

144. D. Hilbert, *Grundlagen der Geometrie*, Leipzig und Berlin, Verlag und Druck von B.G. Teubner, 1930
145. P.H. Hill, *The Science of Engineering Design*, New York, Tufts University, 1970
146. R.A. Hinde, *Animal Behaviour. A Synthesis of Ethology and Comparative Psychology*, New York, McGraw-Hill Book Company, 1970
147. E. Husserl, *Gesammelte Werke*, Bd. VI, Martinus Nijhoff, The Hague, 1954
148. L. Jánossy, *Theory of Relativity Based on Physical Reality*, Budapest, Akad. Kiadó, 1971
149. K. Jaspers, *Die Philosophische Glaube*, München, Piper, 1948
150. J.C. Jones, *Design Methods. Seeds of Human Futures*, London, Wiley-Interscience, 1970
151. I. Kant, *Critique of Pure Reason*, London, G. Bell and Sons, 1930
152. S. Kierkegaard, *Gesammelte Werke*, Abt. 16, Düsseldorf-Köln, Eugen Diederichs Verlag, 1957
153. S.C. Kleene, *Introduction to Metamathematics*, 1952
154. F. Klein, *Elementarmathematik vom höheren Standpunkte aus*, Bd. I, Berlin, Verlag von Julius Springer, 1933
155. F. Klix, *Erwachendes Denken. Eine Entwicklungsgeschichte der menschlichen Intelligenz*, Berlin, Deutscher Verlag der Wissenschaften, 1980
156. A. Koyré, *Etudes Calilèennes*, Vol. 1, *A l'Auber de la science classique*, Paris, Hermann, 1939
157. E.V. Krick, *Methods Engineering, Design and Measurement of Work Methods*, New York, Wiley, 1962
158. T.S. Kuhn, *The Structure of Scientific Revolutions*, The Univ. of Chicago Press, 1962
159. R.E. Leakey, "Skull 1470", *National Geographic*, Washington, 1973, Vol. 143, No. 6
160. G.W. Leibnitz, *Vernunftprinzipien der Natur und der Gnade. Monadologie*, Verlag von Felix Meiner, Hamburg, 1956
161. C.I. Lewis, *A Survey of Symbolic Logic*, New York, Dover, 1960
162. D.-E. Liebscher, *Relativitätstheorie mit Zirkel und Lineal*, Berlin, Akademie-Verlag, 1977
163. J. Lips, *The Origin of Things*, New York, A.A. Wyn, Inc., 1947
164. T. Lipps, *Grundzüge der Logik*, Verlag von Leopold Voss, Hamburg, 1893
165. A. Lowinger, *The Methodology of Pierre Duhem*, New York, Columbia University Press, 1941
166. B. Malinowski, *A Scientific Theory of Culture and Other Essays*, Chapel Hill, The University of North Carolina Press, 1944
167. D. Meister, *Behavioural Foundations of System Development*, New York, John Wiley and Sons, 1976
168. E. Meyerson, *Identitè et Realitè*, Librairie Félix Alcan, Paris, 1926
169. A Miele, *Flight Mechanics*. Vol. 1, "Theory of Flight Paths", London, Palo Alto, 1962
170. J.S. Mill, *A System of Logic Ratiocinative and Inductive*, London, Longmans, Green, and Co., 1900
171. C. Mitcham, "Philosophy of Technology", *A Guide to the Culture of Science, Technology, and Medicine*, New York, the Free Press, 1980
172. C. Mitcham and R. Mackey (eds.), *Philosophy and Technology*, New York, the Free Press, 1972
173. A. Mostowski, *Thirty Years of Foundational Studies*, Helsinki, Soc. Philosophica Fennica, 1965

174. J. von Neumann, "The Mathematician", *Collected Works*, Vol. 1, Oxford, Pergamon Press, 1961
175. E. Olszewski, "The Notion of Technological Research and Its Place among Other Activities", *Polish Essays in the Philosophy of the Natural Sciences*, Vol. 68, Dordrecht-Boston-London, D. Reidel Publishing Company, 1982
176. J. Ortega y Gasset, *The Dehumanization of Art*, New York, Doubleday and Company, 1956
177. J. Ortega y Gasset, *The Modern Theme*, New York, Harper and Row, 1961
178. M. Planck, *Physikalische Abhandlungen und Vorträge*, Bd. III, Braunschweig, Friedr. Vieweg und Sohn, 1958
179. W. Quine, *From a Logical Point of View*, New York, Harper and Row, 1963
180. H. Rickert, *Kulturwissenschaft und Naturwissenschaft*, Freiburg, 1899
181. P. Ricoeur, "Human Sciences and Hermeneutical Method: Meaningful Action Considered as a Text", *Explorations in Phenomenology*, The Hague, Martinus Nijhoff, 1973
182. W. Sellars, *Science, Perception and Reality*, Routledge and Kegan Paul, London, 1963
183. H.A. Simon, *The Sciences of the Artificial*, London, the MIT Press, 1981
184. W.T. Singleton, *Introduction to Ergonomics*, Geneva, WHO, 1972
185. T. Slama-Cazacu, "The Place of Applied Linguistics in the System of Sciences: AL in Relation to 'Linguistics' ", *Revue Roumaine de Linguistique*, 1980, t. XXV, No. 3
186. R.R. Stoll, *Sets, Logic and Axiomatic Theories*, San Francisco, W.H. Freeman and Company, 1961
187. L. Thorndike, *The History of Medieval Europe*, Houghton Mifflin Co., Boston, 1928
188. L. Tondl, "On the Concepts of 'Technology' and 'Technological Sciences' ", *Contributions to a Philosophy of Technology*, Boston, 1974
189. *The New Testament. The Gospel According to S. John*, Cambridge, London, Cambridge University Press, 1903
190. *The Renaissance Philosophy of Man*, Chicago, the University of Chicago Press, 1948
191. *The Search for the Human Sciences*, Deakin University, 1980
192. J.A. Wheeler, "From Relativity to Mutability", *The Physicist's Conception of Nature*, Dordrecht-Boston, D. Reidel Publishing Company, 1973
193. E. Wigner, *Symmetries and Reflections*, Bloomington-London, Indiana University Press, 1967
194. I.G. Wilson and M.E. Wilson, *Management, Innovation and System Design*, Princeton Auerbach Publishers, 1971
195. J.H. Woodger, *The Axiomatic Method in Biology*, Cambridge, The University Press, 1937
196. E. Zilsel, "The Sociological Roots of Science", *The American Journal of Sociology*, Chicago, 1942, Vol. 47, No. 4

NAME INDEX

SUBJECT INDEX